Introduction to Modeling Sustainable Development in Business Processes

Dirk Inghels

Introduction to Modeling Sustainable Development in Business Processes

Theory and Case Studies

Springer

Dirk Inghels
Vrije Universiteit Amsterdam
Amsterdam, The Netherlands

ISBN 978-3-030-58421-4   ISBN 978-3-030-58422-1 (eBook)
https://doi.org/10.1007/978-3-030-58422-1

This Springer imprint is published by the registered company Springer Nature Switzerland AG
The registered company address is: Gewerbestrasse 11, 6330 Cham, Switzerland

Earth provides enough to satisfy every man's needs, but not every man's greed

Mahatma Gandhi

Preface

We are living in a challenging moment in history. Many changes are taking place, including growing digitalization, deglobalization, 3D printing, and decreasing extreme poverty. A major contemporary challenge is dealing with the consequences of global warming and the depletion of natural resources that are linked to what is called the linear economy, also known as the "take-make-dump" economy. A transition towards a circular economy based on reuse, disassembly, and refurbishment or recycling should limit and ultimately stop the extraction of non-renewable natural resources. The speed at which this transition takes place will determine the future viability of planet Earth and the living conditions of future generations.

Sustainable development and corporate social responsibility drive countries, regions, and businesses to take environmental and social concerns into account when realizing economic objectives. A growing awareness of the connectedness between industrial, societal, and environmental systems might shift the way businesses will be operated. This requires taking economic, ecological, and societal objectives into account simultaneously. Moreover, it requires one to evaluate the dynamics of business systems over time and assess the long-term effects on people, the planet, and profits when proposing changes that may support the sustainability transition.

This book aims to help students and business practitioners use quantitative modeling in their pursuit to make business processes sustainable. Two approaches are introduced. In Chaps. 2, 3, 4, 5, and 6, we discuss a linear optimization approach to the three sustainability objectives: economic, environmental, and societal objectives. This requires an introduction to quantifying the three different objectives. Since a sustainable approach to business processes is often characterized by non-linearities and feedback loops, the system dynamics approach can be a suitable modeling approach. This approach is discussed in Chaps. 7, 8, 9, 10, and 12.

I welcome all feedback and suggestions to make this book more relevant and useful.

Writing this book has been a challenge; it required spending time that I was not able to give to my partner Lieve, family, and friends. I'm very thankful for their patience, understanding, and support. I'm also grateful to Wout Dullaert of the Vrije Universiteit Amsterdam, who has actively supported me in my journey to discover how to model sustainability in business process environments. Also, a special thanks to Linda Weix, who reviewed this book and checked the grammar and punctuation.

I hope you enjoy reading the book!

Amsterdam, The Netherlands Dirk Inghels

Contents

1 Sustainability Transition for Businesses 1
1 Emergence of Sustainable Development 2
1.1 Economic Growth from a Historical Perspective 2
1.2 World Population Growth from a Historical Perspective 5
1.3 Sustainable Development as an Answer to the Multiple Crises of Modern Society 7
1.4 Key Pillars of Sustainability 8
2 The Tragedy of the Commons 10
2.1 Ways to Counteract 11
3 Emergence of Sustainable Development in a Business Context 15
3.1 Sustainability in Business Processes 15
4 Towards Sustainable Supply Chain Management 19
4.1 Supply Chain Management 19
4.2 Sustainable Supply Chain Management 20
5 Modeling Sustainable Business Processes 22
6 Case Study: Green Waste Valorization 23
6.1 The Basic System and Its Notation 24
6.2 First Environmental Objective: Maximizing Composting Yield 26
6.3 Second Environmental Objective Function: Maximizing Waste to Energy 26
6.4 Third Environmental Objective Function: Minimizing the Environmental Impact 27
6.5 Fourth Environmental Objective Function: Minimizing the PM Emissions 27
6.6 Societal Impact 28
6.7 Economic Impact 28
6.8 Constraints 29

2 Single Objective Optimization . . . 31
1 The Municipal Solid Waste Allocation Problem . . . 31
2 Standard Form of the Mathematical LP Problem . . . 33
2.1 Generic Form of a Linear Programming Problem . . . 33
2.2 The MSW Allocation Problem Formulated as a Linear Programming Problem . . . 34
3 Assumptions of Linear Programming . . . 36
3.1 Proportionality . . . 37
3.2 Additivity . . . 37
3.3 Divisibility . . . 37
3.4 Certainty . . . 38
3.5 Data Availability . . . 38
4 Determining the Optimal Solution . . . 38
4.1 Graphical Method . . . 38
4.2 Simplex Method . . . 40
4.3 Algebraic Method . . . 41
4.4 Using the MS Excel Solver . . . 45
5 Special Cases . . . 51
6 Optimizing Profit for the Green Waste Valorization Case . . . 52
7 Mixed Integer Linear Programming (MILP) . . . 55
7.1 Generic Form of a MILP Problem . . . 55
7.2 The Adapted MSW Allocation Problem Formulated as a MILP . . . 56

3 Multiple Objective Optimization . . . 59
1 Introduction: Bi-objective Optimization – The Green Waste Valorization Case . . . 59
2 The Difference Between a SOOP and a MOOP . . . 62
3 The Standard Form of a MOOP . . . 63
4 Two Approaches to Obtain One Single Solution in Multi-objective Optimization . . . 64
5 The Preference-Based Procedure . . . 65
6 The Ideal Procedure . . . 67
6.1 Step 1- Determining the Pareto-optimal Front . . . 67
6.2 Step 2- Selecting a Single Optimal Solution on the Pareto-optimal Front . . . 74
7 The Augmented ε-Constraint Method . . . 75
8 Other Solution Methods . . . 79

4 Quantifying the Economic Impact . . . 81
1 Quantification of Economic Impact in a Business Context and Supply Chains . . . 81
2 Cost, Profit and Cash Flow . . . 83
2.1 Cost . . . 83
2.2 Profit . . . 83
2.3 Cash Flow . . . 83
2.4 Single-Payment Compound-Amount Factor . . . 84
2.5 Equal-Payment-Series Compound-Amount Factor . . . 86
3 Internalizing External Costs . . . 88

5 Quantification of the Environmental Impact . . . 89
1 Introduction . . . 89
1.1 The Generic Life Cycle of a Product . . . 89
1.2 The Purpose of Life Cycle Assessment . . . 92
2 Commonly Used Terminology . . . 93
3 The Four-Stage Approach of an LCA According to ISO 14040+44 . . . 94
3.1 Goal and Scope Definition . . . 95
3.2 Life Cycle Inventory Analysis (LCI) . . . 101
3.3 Life Cycle Impact Assessment (LCIA) . . . 102
3.4 Interpretation . . . 103
4 LCA Applied to the Green Waste Valorization Case Study . . . 103
4.1 Goal and Scope . . . 104
4.2 Life Cycle Inventory . . . 105
4.3 Life Cycle Impact Assessment . . . 106
4.4 Interpretation . . . 107
5 LCA Used as an Input in MCDA . . . 107

6 Quantifying the Social Impact . . . 109
1 Introduction to the AHP . . . 109
1.1 Pairwise Comparison of Criteria and Alternatives . . . 110
1.2 Generalization . . . 115
2 Computing the Priority Vector . . . 117
2.1 The Eigenvector Method . . . 117
2.2 The Geometric Mean Method . . . 119
3 AHP Applied to the Social Impact of the Green Waste Valorization Case . . . 120
3.1 The Eigenvector Method . . . 124
3.2 The Geometric Mean Method . . . 128
4 The Approximation Method . . . 129
5 AHP Using Excel . . . 131
5.1 Calculation of the Priority Vectors Using the Geometric Mean Method . . . 133
5.2 Calculation of the Priority Vectors Using the Approximation Method . . . 135

7 Systems Thinking and Introduction to System Dynamics Modeling 141
1 Systems Thinking: Challenging the Cartesian-Newtonian Paradigm 143
2 The Paradigm of Economic Growth 144
3 Introduction to System Dynamics Modeling 146

8 Causal Loop Diagrams 149
1 Causal Loop Diagram Notation 149
2 Practical Use of Causal Loop Diagrams 151
2.1 Avoid Spurious Correlations 151
2.2 Determining Loop Polarity 151
2.3 Depicting Important Delays in Causal Links 153
2.4 Variable Names in Causal Diagrams 153

9 Structure and Fundamental Modes of Behavior in Dynamic Systems 155
1 Basic Modes of Dynamic Behavior 155
1.1 Exponential Growth 155
1.2 Goal Seeking 158
1.3 Oscillation 161
2 Interaction of the Basic Modes 161
2.1 S-Shaped Growth 162
2.2 S-Shaped Growth with Overshoot 166
2.3 Overshoot and Collapse 166
2.4 Tragedy of the Commons 167
2.5 Other Modes of Behavior 168

10 Stocks and Flows and the Dynamics of Simple Structures 169
1 Introduction 169
2 Variables and Model Boundaries 171
3 Dynamics of Exponential Growth 172
4 Dynamics of Goal-Seeking Behavior 176
5 Dynamics of S-Shaped Growth 181

11 Delays 185
1 Introduction 185
2 Pipeline (Material) Delay 185
3 First-Order Material Delay 189
4 N-th Order Material Delay 198
5 Information Delays 199
6 Dynamics of Oscillating Behavior 202
7 Dynamics of Limits to Growth 206

12 Nonlinear Behavior 215
1 Table Function 215
2 MIN and MAX 216
3 IF. . .Then, Else 216
4 Caste Study: My Domestic Power Consumption 216

References 225

Chapter 1
Sustainability Transition for Businesses

In this initial chapter, we discuss the main drivers behind the sustainability transition that many businesses are currently facing.

Since the start of the Industrial Revolution, the global economy has experienced unprecedented economic growth. The average world gross domestic product (GDP) per capita increased by a factor of 10 between 1820 and 2010. However, this rise has been unequal for different parts of the world (Bolt et al. 2014).

An increasing number of signs indicate that this growth is no longer sustainable because it damages the environment and society. Global warming and the speedy depletion of natural resources are two well-known consequences associated with how economic growth has been created. The long-lasting overshoot of the global environmental footprint forces societies to look differently at economic growth on a macroeconomic level, based on a connection with the environment and society. A growing global population reinforces the consequences of unsustainable economic growth.

On a micro-economic level, businesses can contribute to a sustainable society by applying the insights of environmentally-conscious manufacturing and converting their operations into sustainable operations. This approach takes the life cycle of products into account, from the design phase to product manufacturing, product use, and product recovery. Other approaches, such as a reduction of waste, bolstered by the insights of lean management, help make businesses more sustainable as well.

The transition towards a sustainable society, sustainable business processes, and sustainable development is generally hindered by short-term perspectives. Those who benefit most from maintaining the status quo have little reason to promote change. We will discuss the underlying dynamics of this phenomenon, called 'the tragedy of the commons,' to explain how to shift from a focus on individual gain towards collective gain by imposing taxes and other financial corrective measures.

We will end this chapter by presenting a case study on Green Waste Valorization (GWV) that will be used throughout the next five chapters as the primary,

D. Inghels, *Introduction to Modeling Sustainable Development in Business Processes*, https://doi.org/10.1007/978-3-030-58422-1_1

overarching example demonstrating the quantitative modeling techniques that are presented in the subsequent chapters.

1 Emergence of Sustainable Development

Two major reinforcing processes have fueled the current global overshoot in natural resource use and the associated generation of excessive waste and pollution: ever-increasing consumption per capita coupled with constant population growth. We discuss these two phenomena separately before discussing how sustainable development can be a solution to balancing these reinforcing processes.

1.1 Economic Growth from a Historical Perspective

Countries and regions usually express their economic growth as the percentage increase in real Gross Domestic Product (GDP). This is an increase in the inflation-adjusted market value of the goods and services produced by an economy over time. It can be individualized by expressing GDP in terms of income growth per capita, which triggers consumption per capita. Economic growth is considered positive because it is believed to drive an improvement in the quality of life. In contrast to the last two previous centuries, economic growth and the associated improvements in peoples' standards of living have increased relatively slowly for most of human history.

Until the first half of the seventeenth century, the growth in income per person was related to population size. Population growth led to a decline in nutrition and income per person. This is known as the *Malthusian Trap*. Technological breakthroughs before the first half of the seventeenth century, such as the invention of windmills and irrigation technology, led to a temporary rise in income per person, followed by a permanent increase in population and a decline in income per person. In other words, higher productivity resulted in larger, but not wealthier populations.

Starting around 1650, new inventions related to the Industrial Revolution initiated economic growth in England and Holland, ushered in a new era of unprecedented and continuous growth of incomes, and led to the decoupling of income growth and population size. This new era of continuous and sustained economic growth gradually spread to many other countries (see Fig. 1.1). Economic growth has resulted in widespread global prosperity, better nutrition, longer life expectancy, earlier retirement, more urbanization, and better healthcare for many people (Roser 2020).

In recent decades, environmental and social concerns have questioned the viability and purpose of long-lasting, continuous economic growth. Income equality has been a growing social theme in developed countries since the 1980s (Piketty 2014). Those in a lower income class benefit less from economic growth, and sometimes they balance on the frontier of poverty. From an environmental point of view, economic growth has been connected to the depletion of our natural capital and

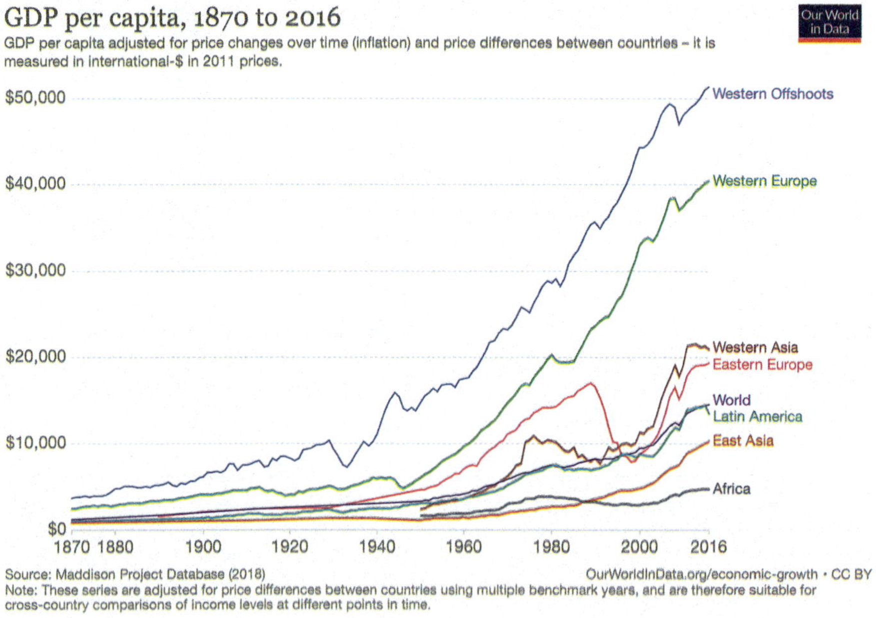

Fig. 1.1 Evolution of world GDP. (Source: ourworldindata.org/economic-growth)

the earth's declining regenerative capacity since the early 1970s. This is reflected in the expectation that the era of cheap and abundant oil, upon which economic expansion was built, is over (Bonaiuti 2014; Rifkin 2010). Scarcity is soon expected for many primary resources because we are extracting resources faster than our planet can regenerate them. Moreover, large quantities of carbon emissions and global warming are related to the use of non-renewable, carbon-based fuels needed for economic growth.

We must counteract the overuse of natural resources. A rapid reduction in our ecological footprint is needed if we are to prevent the ultimate collapse of our natural ecosystem. Figure 1.2 depicts the historic and forecasted global ecological footprint evolution from 1960 to 2030. The global Ecological Footprint Calculator expresses the current overshoot in the sustainable use of natural resources (www.footprintnetwork.org). Starting in 1970, with an initial global average ecological deficit of more than one Earth, this figure has been increasing towards an almost constant ecological deficit of 1.75 Earths between 2011 and 2019 (Global Footprint Network 2019). If we are not able to reduce this ecological deficit below one Earth over the next decades, an environmental and economic collapse cannot be avoided (Meadows et al. 1972). Failure to take planetary boundaries into account will drive the earth to a less environmentally viable state, which could have a significant impact on our society (Stockholm Resilience Centre 2020).

Finite planetary resources force us to look at the relation between economy, environment, and society differently. The current paradigm that GDP growth is needed to keep or even improve our welfare and wellbeing in the West (North

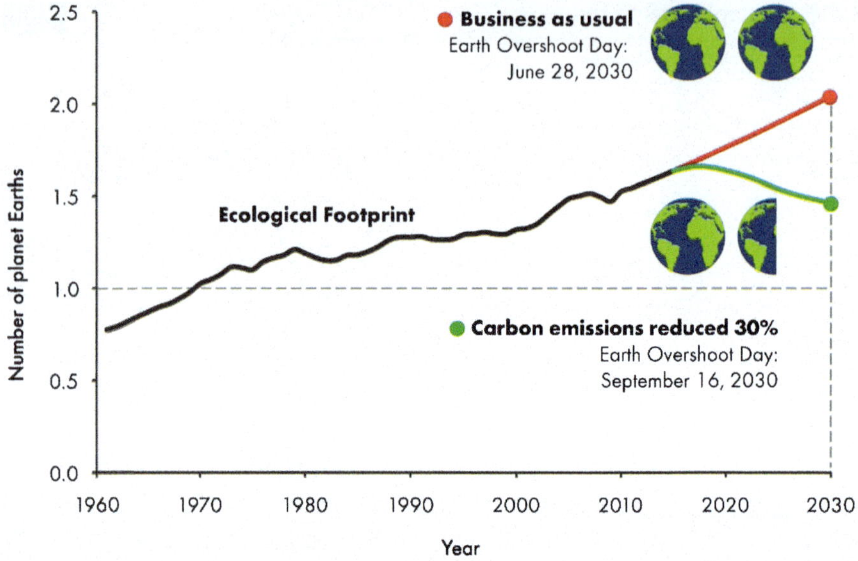

Fig. 1.2 Evolution and forecast of Ecological Footprint expressed in number of Earths needed. (Source: Global Footprint Network 2019)

America and the EU zone) could be questioned. In this respect, it is worth considering the environmental economists' point of view on the limits of economic growth: Manufactured goods and services are consumed to meet the satisfaction level of people's needs and desires, also called *utility*. To produce these goods and services, sacrifices must be made; these include the use of labor, resource depletion, pollution exposure, etc., also called *disutility*. The balance between utility and disutility when producing and consuming an increased number of manufactured goods and services determines whether growth can still be called economic growth, according to Daly (2014, 2008), as depicted in Fig. 1.3. If the marginal utility, i.e., the utility added by consuming one more unit of goods and services, outweighs the marginal disutility, then we speak of economic growth up to the "*economic limit*", according to Daly. The economic limit is defined by marginal cost equal to marginal benefit and the consequent maximization of net profit. It is the point in Fig. 1.3 where marginal utility equals marginal disutility.

Further growth beyond this point is uneconomic growth since marginal disutility outweighs marginal utility. In the region of uneconomic growth, two types of limits can be distinguished. The first is the "*futility limit*" or the limit when the marginal utility of production falls to zero, i.e., more consumption is no longer beneficial. The second limit of uneconomic growth is the "*ecological catastrophe limit*" or the point in growth where a chain reaction or tipping point is induced, leading to the collapse of our ecosystem. Global warming caused by anthropomorphic greenhouse gas emissions could induce such a tipping point (IPCC 2020).

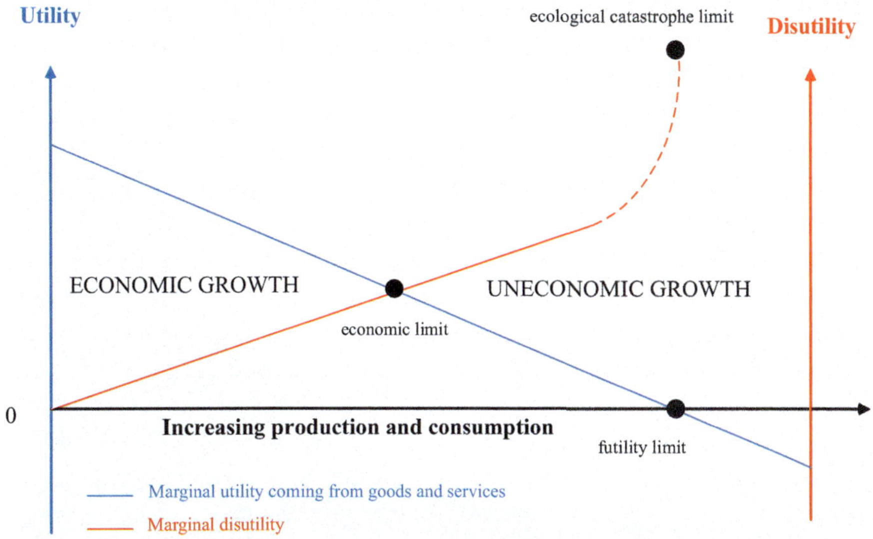

Fig. 1.3 The three limits of Economic Growth according to Daly. (Based on Daly 2014)

Increasing production and consumption is rightly called economic growth up to the economic limit. However, any increase in real GDP is commonly taken into account for economic growth even if it increases costs (production costs as well as environmental and societal costs) faster than benefits. According to Daly, adopting a steady state economy will allow us to avoid surpassing the economic limit. A steady state economy entails a stabilized population (keeping birth rates on par with death rates) and stabilized per capita consumption (keeping production rates on par with depreciation rates). Minimizing waste at higher levels of production and consumption allows for a steady state economy.[1] Similar to Daly (2014, 2008), Meadows et al. (1972) concluded that a steady state economy is the ultimate solution to safeguarding the environmental viability of the planet.

1.2 World Population Growth from a Historical Perspective

The environmental decline linked to unsustainable economic growth is reinforced by a growing global population. In 2019 there were about 7.7 billion people on Earth. The UN forecasts that the world population will grow to 10.9 billion by 2100. Population growth is determined by the number of births versus deaths. If the global birth rate exceeds the global death rate, then the world population increases. Figure 1.4 shows that we are currently facing a trend of decreasing population growth

[1] see: https://steadystate.org/discover/definition/

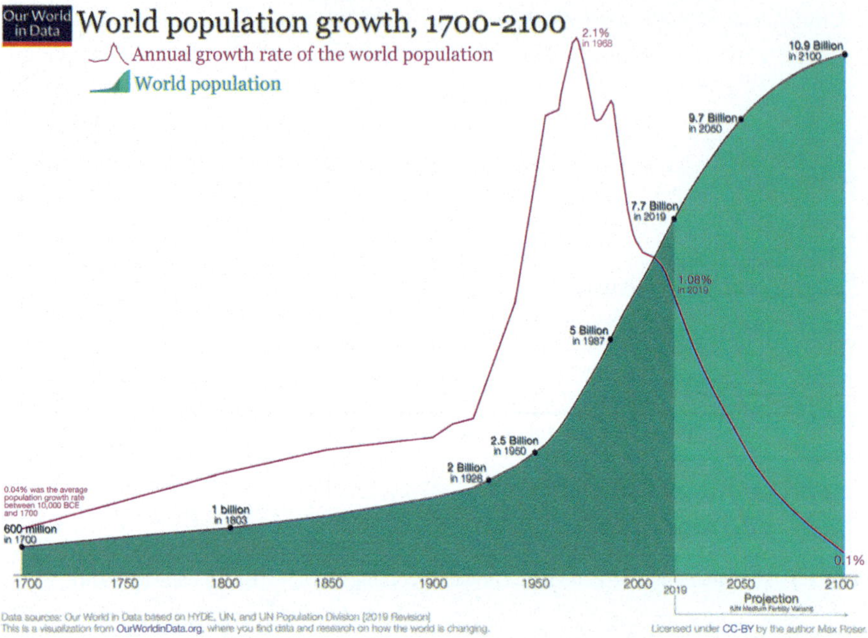

Fig. 1.4 Evolution and forecast of world population growth. (Source: https://ourworldindata.org/future-population-growth#projections-of-the-drivers-of-population-growth)

rate that is expected to last until the end of this century. The world population explosion is likely to come to an end because many countries have already passed or will assumedly pass the demographic transition in the near future.

The theory of demographic transition involves four phases whereby societies evolve from a high birth and death rate towards a low birth and death rate. In the first phase, societies evolve following the Malthusian paradigm, i.e., the population size is inversely proportional to the food supply. The total cost of raising children in this stage is on par with the financial contribution to the household. When societies start to develop, life expectancies increase because of improvements in the food supply and a reduction in diseases. This second phase of the demographic transition still has high birth rates but lower death rates. Consequently, the number of people in society rises. In the third stage, population growth begins to level off due to a decline in the birth rate. This decline is triggered by a number of factors linked to the development of society, such as higher wages, higher levels of education, the changed position of women who also gain access to education, a reduction in child labor, access to contraception, etc. Finally, in the fourth stage, both birth and death rates are low, and birth rates may even drop below replacement level.

It is not surprising that the first decline in birth rates in developed countries started in the late nineteenth century in northern Europe, where the Industrial Revolution started. Since then, many other countries and regions have followed this trend. However, a significant number of countries are still in phase one of the demographic transition. Many of them are in Africa.

1.3 Sustainable Development as an Answer to the Multiple Crises of Modern Society

Continuous economic growth, combined with a continually growing population, are two reinforcing processes that drive the depletion of natural resources in order to satisfy global human consumption. Figure 1.5 depicts the situation in a linear, organized economy that is characterized by unrestricted depletion of natural resources and unlimited production of waste, emissions to land, water and air, and unlimited land use. As a consequence, natural sinks like oceans and the atmosphere are no longer able to provide the necessary regeneration of our ecosystem. Many natural sources will eventually be depleted.

In a state of environmental overuse, the distribution of societal wealth can no longer solely be linked to material growth since creating more wealth by increasing production (the traditional solution) could potentially destroy our ecosystem and harm future generations. On the other hand, current generations in developing countries will be harmed if we do not increase our consumption. This is called the *paradox of sustainable development* (Rees 1988).

Sustainable development appeared on the global agenda as an answer to the combination of environmental degradation, lasting poverty, and underdevelopment. The transition towards a sustainable society began with the report "*Our Common*

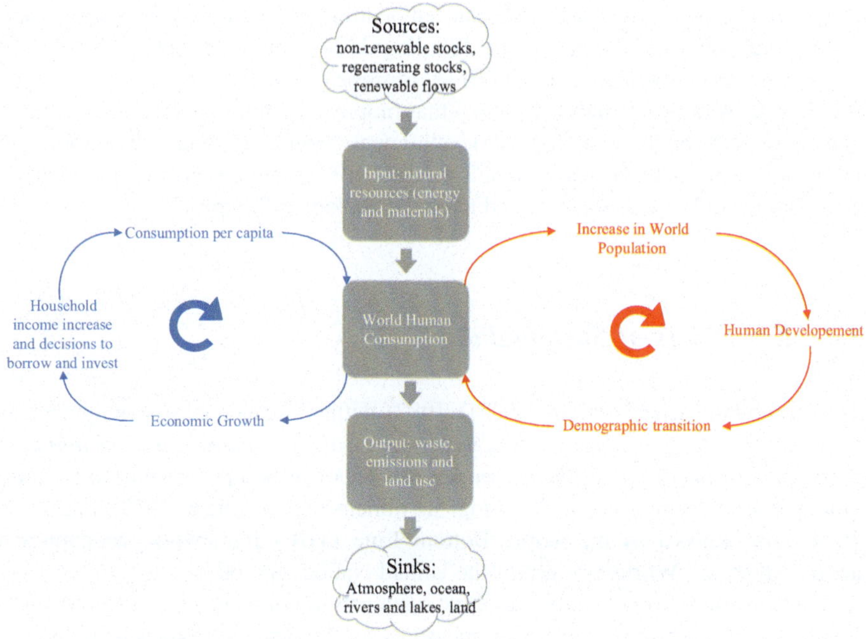

Fig. 1.5 Economic and population growth as two reinforcing processes causing a growing environmental footprint

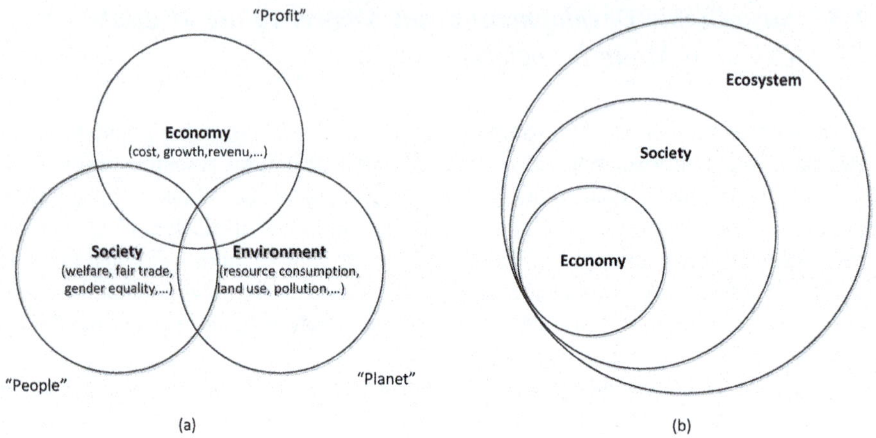

Fig. 1.6 Triple bottom line weak sustainability perspective (**a**) and strong sustainability perspective (**b**)

Future" (also known as 'the Brundtland report') (WCED 1987), and with "*Agenda 21*" of the United Nations Conference on Environment and Development in Rio de Janeiro in 1992 (UN 1992). This transition requires a shift in root perspective towards an ecological paradigm, or worldview, based on finite biophysical limits of the earth and the interdependence of processes (Peeters 2015).

Environmental economists advocate embedding the economy in society, and embedding society in the ecosystem (Sterling 2004). The latter approach is called *strong sustainability* (Fig. 1.6b). It differs, to some extent, from *weak sustainability* (Fig. 1.6a). Weak sustainability claims that biophysical limits need to be explored and enlarged through technological development. Strong sustainability is rooted in the idea of "limits to growth" (Meadows et al. 1972) and claims that the planet's ecosystem should be restored by reducing the ecological footprint.

1.4 Key Pillars of Sustainability

The Brundtland Commission (WCED 1987) defined sustainable development as "*development that meets the needs of the present without compromising the ability of future generations to meet their own needs.*" This can be operationalized by integrating social, economic, and ecological dimensions (Hediger 1999; Elkington 1994), also known as the Triple Bottom Line (TBL) framework, an approach advocated by the World Summit of the United Nations in 2005.

The TBL framework requires all three sustainability dimensions to be quantified. Figure 1.7 summarizes some typical measures for the three sustainability pillars.

The economic pillar is commonly represented by the minimization of costs or the maximization of profits (Seuring and Müller 2008). Since the World Summit on

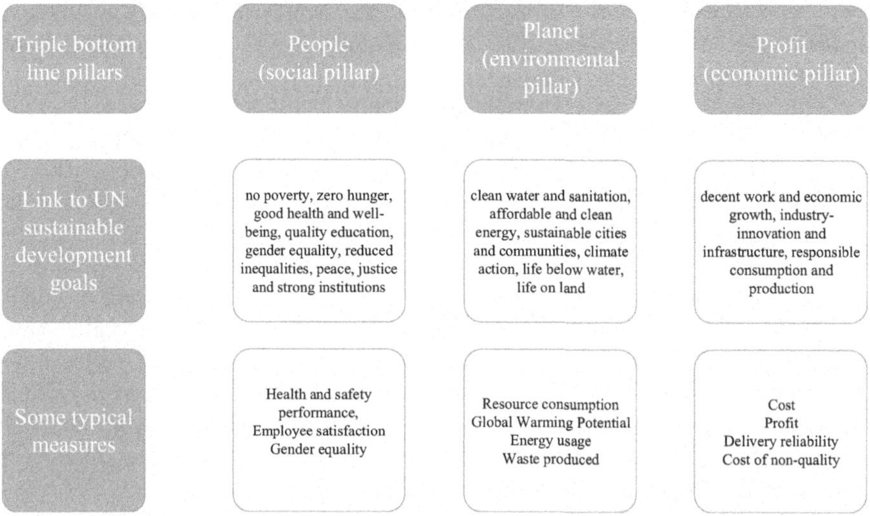

Fig. 1.7 People-Planet-Profit framework introduced at the 2005 World Summit of the United Nations

Sustainable Development in Johannesburg in 2002, prosperity has been used, instead of profit, to reflect the perspective that the economic dimension covers more than company profits (Heijungs et al. 2010).

The standard assessment tool for the environmental pillar is the Life Cycle Assessment (LCA), described in ISO 14040 + 44 (2006), which determines the life cycle impact of processes and products. The Eco-indicator methodology (Goedkoop and Spriensma 2001) can also be applied to quantify environmental impacts in a simplified manner. As the economic and environmental dimensions have been on the agenda for some time, there is a growing consensus on how to describe them (Seuring and Müller 2008), and several (optimization) models can be found in the academic literature.

A commonly accepted definition is not yet available for the social dimension, mainly because there is no consensus on the meaning of the term 'social' (Lehtonen 2004). The social dimension is immaterial and, therefore, difficult to analyze quantitatively (Lethonen 2004; Munda 2004). Since many social indicators cannot be quantified, qualitative ranking and scoring are currently used alongside quantitative approaches (Klöpffer 2008). A popular Multi-Criteria Decision Making (MCDM) method that can be used to quantify such qualitative comparisons is the Analytic Hierarchy Process (AHP) (Saaty 2008). The AHP requires that the social criterion of interest be selected and rated using pairwise comparisons, as illustrated in Dehghanian and Mansour (2009). They used the AHP to compare social impacts for several design options to obtain a sustainable recovery network for end-of-life tires in Iran.

Considering supply chain management as one of the most important business processes, Seuring (2013) and Sharma et al. (2013) concluded that only a limited number of papers on green or sustainable (forward) supply chains apply quantitative models. Moreover, the social aspect of sustainability is often ignored in these quantitative models.

In the transition towards a sustainable society, investments have to be made to realize environmental improvements in products, processes, infrastructure, and transportation. Societal choices will also have to be made. For example, take the transition towards low carbon transportation. It requires investments in environmentally-friendly alternatives such as electric vehicles, which are more expensive than traditional fossil fuel-powered vehicles. Clothing made in countries where people are not exploited will be more expensive, but is the consumer willing to pay more? Many people will, in general, share the environmental and societal benefits associated with more sustainable alternatives where costs may be local to a firm or individual. However, keeping the status quo will often provide more profit in the short term for firms and individuals, while the burden and costs associated with environmental and societal consequences will be allocated globally. This dilemma is described in *"The Tragedy of the Commons"* (Hardin 1968).

2 The Tragedy of the Commons

Biology professor Garret Hardin wrote an influential article in Science entitled "*The Tragedy of the Commons*" (Hardin 1968). The subtitle he added was: "The population problem has no technical solution; it requires a fundamental extension in morality." In this article, he advocates some ideas that are important to sustainable development. Since the population tends to grow exponentially, the per capita share of the world's goods must steadily decrease since we are living in a world with finite resources. A finite world can only support a finite population. Consequently, population growth must eventually equal zero. Hardin directly links our desire for continuously prosperous growth: "*there is no prosperous population in the world today that has, and has had for some time, a growth rate of zero. Any people that has intuitively identified its optimum point will soon reach it, after which its growth becomes and remains zero*".

Adam Smith described the link between individual happiness and prosperity in his publication "*The Wealth of Nations*" (Smith 1776). Smith promoted the theory of "the invisible hand," i.e., he believed that each individual intends to maximize its own gain but, at the same time, is led by an invisible hand to promote the public interest. Hardin concludes that the tendency to assume that decisions reached individually will, in fact, be the best decisions for an entire society has been widely accepted since Adam Smith's publication. He immediately continues that if Smith's assumption is not correct, we need to re-examine our individual freedoms to see which ones are defensible.

To discuss the rebuttal of the invisible hand in population control, he developed "the tragedy of the commons" as follows: "*Picture a pasture open to all. It is to be expected that each herdsman will try to keep as many cattle as possible on the commons. Such an arrangement may work reasonably satisfactorily for centuries because tribal wars, poaching, and disease keep the numbers of both man and beast well below the carrying capacity of the land. Finally, however, comes the day when the long-desired goal of social stability becomes a reality. At this point, the inherent logic of the commons remorselessly generates tragedy. As a rational being, each herdsman seeks to maximize his gain. Explicitly or implicitly, more or less consciously, he asks, "What is the utility* **to me** *of adding one more animal to my herd?" This utility has one negative and one positive component. (i) The positive component is a function of the increment of one animal. Since the herdsman receives all the proceeds from the sale of the additional animal, the positive utility is nearly +1. (ii) The negative component is a function of the additional overgrazing created by one more animal. Since, however, all the herdsmen share the effects of overgrazing, the negative utility for any particular decision-making herdsman is only a fraction of −1. Adding together the components partial utilities, the rational herdsman concludes that the only sensible course for him to pursue is to add another animal to the herd. And another; and another. . . . But this is the conclusion reached by each and every rational herdsman sharing a commons. Therein is the tragedy. Each man is locked into a system that compels him to increase his herd without limit in a world that is limited. Ruin is the destination toward which all men rush, each pursuing his own best interest in a society that believes in the freedom of the commons. Freedom in commons brings ruin to all.*"

Many examples are available today: overfishing, abundant pollution, waste creation, etc. Since the surrounding air and waters cannot be fenced, they have to be considered as commons. As long as it costs less for the polluter to pollute than to purifying the waste before it is dumped, the rational polluter will continue to pollute. All these problems are a consequence of population growth, according to Hardin.

2.1 Ways to Counteract

Hardin suggests a few possible solutions to alleviate the root cause of the problem that commons are free. A first solution consists of selling the commons off as private property. Another one is to keep them as public property and to agree on the allocation of entrances that might be based on wealth, by lottery on a first-come, first-served basis, or by the use of an auction system. Next, he proposes mutual coercion, mutually agreed upon by the majority of the people, because he believes that some sort of coercion will be necessary to change the behavior of individuals. Hardin makes the comparison with a bank robbery: a bank robber considers the bank to be a commons when he takes money from it. Since almost all of us agree that a bank should not be considered a commons, bank-robbing is considered immoral. Mutual coercion can be realized through a *command-and-control* approach or through *market mechanisms* (Chopra and Meindl 2015).

In a command-and-control approach, standards are set by an overarching authority, and all participants have to adhere to them. Many EU directives such as the

Waste Directive 2008/98/EC (EP&C 2008), the Waste Electrical & Electronic Equipment (WEEE) Directive 2003/108/EC (EP&C 2003), and the End-of-Life Vehicle (ELV) Directive 2000/53/EC (EP&C 2000) are examples of the command-and-control approach.

Mutual coercion by market mechanisms attempts to link the price of the commons to the economic activity. This can be put in place using a cap-and-trade mechanism, which limits aggregate emissions by limiting the number of tradable allowances or by imposing emission taxes.

A well-known cap-and-trade example that controls the reduction of greenhouse gases is the *European Union Emissions Trading System* (EU ETS). Launched in 2005, it cuts the emission of carbon dioxide and other greenhouse gases at least cost.[2]

The EU ETS is active in 31 countries, i.e., all 28 EU member countries plus Iceland, Liechtenstein, and Norway. It limits emissions from more than 11,000 installations that use lots of energy, such as power stations and industrial plants, and airlines operating between these countries. In the Netherlands, approximately 400 installations were covered under the EU-ETS in 2012. The current EU ETS covers around 45% of the EU's greenhouse gas emissions. In July 2015, the European Commission presented a legislative proposal to revise the EU ETS for its next phase (2021–2030), in line with the EU's 2030 climate and energy policy framework. The proposal aims to reduce EU ETS emissions by 43% in 2030 compared to 2005.

The 'cap-and-trade' principle of the EU ETS works as follows: a cap is set on the total amount of certain greenhouse gases that can be emitted by installations covered by the system. By reducing the cap over time, total emissions should fall. Within the cap, companies can receive or buy tradable emission allowances. By restricting the total amount of allowances, their value is ensured. If a company is not able to cover all its emissions with its allowance for 1 year, it will be fined. The goal is to set the carbon price high enough to ensure that companies will invest in cleaner technologies. Trading has another advantage: whenever emissions are cut where it costs less to do so, it brings flexibility into the system. The total greenhouse gas emissions decrease, and companies that need more time to reduce their greenhouse gasses can get it.

In the period 2005–2012, CO_2 emissions in the EU decreased by 11.5%, as depicted in Fig. 1.8. This result harbors pluses and minuses. The most significant decrease in CO_2 emissions has been originated by technological changes, leading to an 18.5% decrease for the period under consideration. This decrease is partly outweighed by a growth in economic activity (GDP), which caused a 6.8% increase in CO_2 emissions and a small increase in emissions of 1.7% due to an expansion of the manufacturing sector in Germany and an increase in manufacturing in Eastern European member states.

There are also some drawbacks to the EU ETS. The EU ETS started in 2005 with the distribution of free emission rights for each installation without many restrictions. After a peak in the EU allowance (EUA) price in 2008, just before the economic downturn, the emission allowance market became saturated. However,

[2]see: https://ec.europa.eu/clima/policies/ets_en

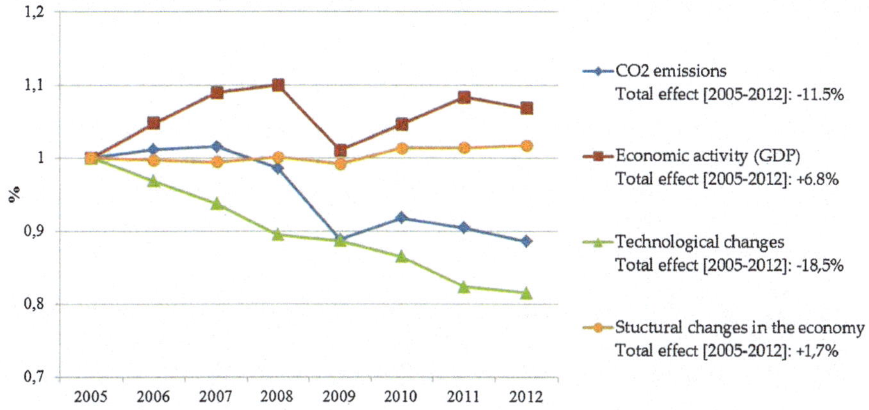

Fig. 1.8 Decomposition analysis of the change in CO_2 emissions from fossil fuel combustion in the EU for the 2005–2012 period. (Source: http://eur-lex.europa.eu/legal-content/EN/ALL/?uri=COM:2015:0576:FIN)

the economic downturn started in 2009 led to a considerable drop in the price per emitted ton of CO_2 in the period 2009–2013. By early 2013, the emission price was historically low (less than €6 per ton), a price level that was not deterrent enough to ensure that CO_2 savings were met by the countries that signed the objectives for 2020. The EUA remained low in the years after 2013 due to a second economic slowdown in Europe. This has undermined the ETS as a driving force towards decarbonization.

The EU ETS finally became a resilient system that supports EU legislation on renewable energy and can cope with future demand shocks in 2018. In 2017, the EU agreed on the Market Stability Reserve (MSR), which withholds surplus EUAs from the market. This helps reduce future oversupplies. The MSR will curb the volume of EUAs annually by 2019. S&P Global Platts (2018) expects that the MSR will better meet the cap-and-trade purposes: "*Higher carbon prices are likely to boost the profitability of companies operating nuclear, wind, solar and hydro-electric power plants, driving further growth in renewable energy capacity in Europe. They also signal a long-term drop in the use of the most emissions-intensive fuels for power generation, hard coal and lignite, and provide a stimulus for innovation in low-carbon industrial goods and processes.*"

Another market mechanism used to enact mutual coercion is imposing an *emission tax*. An example is a carbon tax that governments levy on emitted greenhouse gases. The idea is that producers and/or consumers will take action to switch to lower greenhouse gas emitting alternatives to avoid paying taxes. Contrary to the cap-and-trade mechanism, a carbon tax does not increase when the level of emissions increases. A carbon tax can be levied at any point in the energy supply chain: (i) upstream at the energy producer, (ii) midstream (e.g., power plants), or (iii)

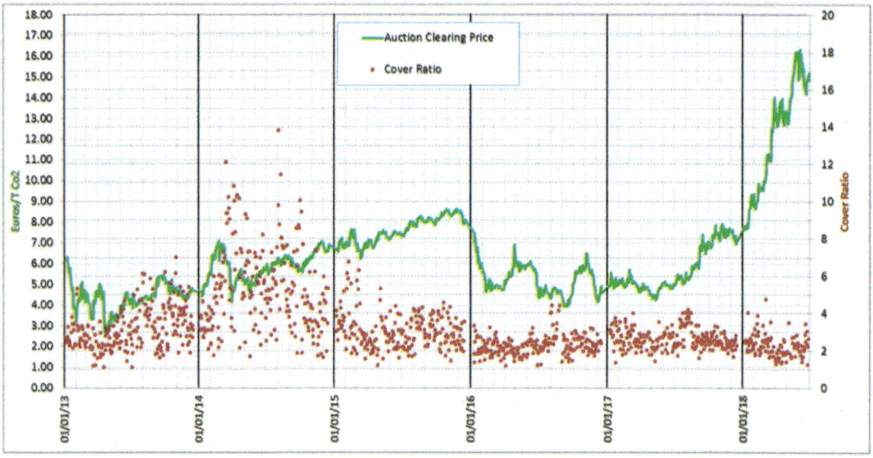

Fig. 1.9 Overview of general allowances auctions from 2013 to 30 June 2018 (Source: European Commission 2018)

downstream at the consumer. Carbon taxes are most effective when they increase over time, forcing producers and consumers to find more greenhouse-friendly solutions. Governments can use carbon tax revenue to reduce taxes on labor or reinvest it in low-carbon investment alternatives. The first carbon tax implementation took place in Finland in 1990, followed by Sweden and Norway in 1991.

Emission taxes, such as carbon taxes, also have drawbacks. Carbon taxes put a higher burden on low-income households. This can cause social unrest like the 'yellow jackets' protests in France. Low-income households have a limited ability to invest in low carbon alternatives that are more expensive than the systems they have to replace. Moreover, emission taxes can hinder competition with countries that do not levy greenhouse emission taxes on the production of goods. Effective implementation also implies that governments stop subsidizing coal, oil, and gas-producing companies, start subsidizing low-carbon investment initiatives such as wind, solar, and hydropower, stimulate energy efficiency, and provide alternatives like cheap and accessible public transport (Fig. 1.9).

Unlike the cap-and-trade mechanism, a carbon tax does not need a monitoring system. A cap-and-trade mechanism offers companies financial incentives to reduce their greenhouse gas emissions actively. However, companies will not take decarbonization improvement actions if carbon credits can be purchased at prices that are less than the cost of shifting to low-carbon solutions. Another drawback of the cap-and-trade mechanism is that the price of the tradable emission rights is unpredictable. This can slow down investment in low-carbon solutions due to the unknown payback period.

3 Emergence of Sustainable Development in a Business Context

In the previous section, we discussed the drivers behind the current unsustainable economic growth, on a macro-economic level, in the sense that it comprises global environmental damage and that it entails societal flaws. In this section, we shift the focus towards sustainability in a business context, and we explore how business processes can contribute to sustainable economic growth taking environmental and societal constraints into account. This is known as *Corporate Social Responsibility*.

3.1 Sustainability in Business Processes

Sustainability in business processes can be defined as the combined economic, environmental, and social optimum of manufacturing alternatives that take into account constraints such as technological limits or legislation. This is also known as the triple bottom line (TBL) approach to People-Planet-Profit optimization (Kleindorfer et al. 2005). Many sustainability assessments are based on the TBL accountancy concept (Seuring 2013), which expands classic financial reporting to include social and environmental performance, as proposed by Elkington (1994). A criticism of the TBL approach is that separating the sustainability concept into three pillars tends to emphasize potentially competing interests between pillars, rather than focusing on their interdependencies (Pope et al. 2004).

Many approaches support making business processes sustainable. In the next sections, we will discuss two of them: *environmentally conscious manufacturing* and *lean manufacturing*. Both approaches support a transition from a linear to a circular economy.

3.1.1 Environmentally Conscious Manufacturing and Product Recovery

Environmentally conscious manufacturing (ECM) deals with green principles for manufacturing products – from conceptual design to final delivery to consumers and, ultimately, to end-of-life (EOL) disposal – that satisfy environmental standards and requirements (Ilgin and Gupta 2010). Environmentally conscious manufacturing and product recovery (ECMPRO) have come to be seen as obligations to the environment and society at large. Primarily, this has been enforced by governmental regulations and driven by customer perspectives on environmental issues (Gungor and Gupta 1999). Figure 1.10 depicts the interactions among the activities that take place in a product life cycle.

In order to counteract the fast depletion of primary raw materials and to decrease the amount of waste, environmentally-friendly products must be developed along

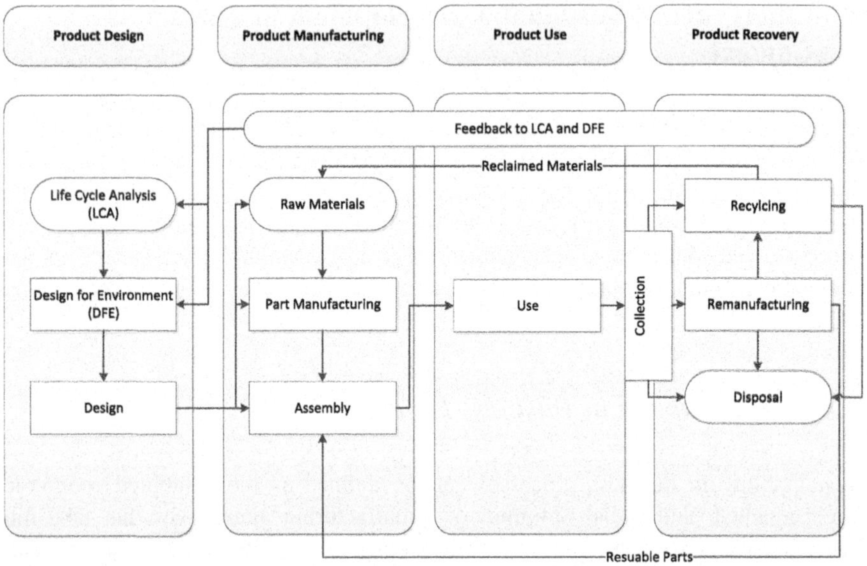

Fig. 1.10 Interactions among the activities in a product life cycle according Gungor and Gupta (1999)

with techniques for product recovery and waste management. In the Product Design phase, Life Cycle Analysis (LCA) can be used to assess the environmental impact of product design, manufacturing, use, and recovery – an approach called "Design for Environment" (DFE). Making better choices in the early stages of product development can have the biggest impact on the environment. In the next phase, Product Manufacturing, the recovery of products should be taken into account to any extent possible. This is generally achieved in two ways: via recycling and remanufacturing. Recycling aims to recover raw materials so that they can be used in new products. Remanufacturing aims to bring old products back to an as-new level of quality. After their useful life, products are discarded and become waste. The waste hierarchy described in the European Waste Directive 2008/98/EC (EP&C 2008) advocates prevention above reuse, recycling, and incineration with energy recuperation. Products should only be disposed of if no other valorization option is (currently) possible. Both manufacturers and consumers are forced to pay more attention to environmental issues related to products as a result of new laws, legislation, and taxation. Examples in Europe include the Waste Directive and the End-of-Life Vehicle (ELV) Directive 2000/53/EC (EP&C 2009).

The literature on ECMPRO is organized into four main areas: product design, reverse and closed-loop supply chains, remanufacturing, and disassembly. This is depicted in Fig. 1.11 (Ilgin and Gupta 2010). For a comprehensive literature survey on ECMPRO we refer to Gungor and Gupta (1999) and Ilgin and Gupta (2010).

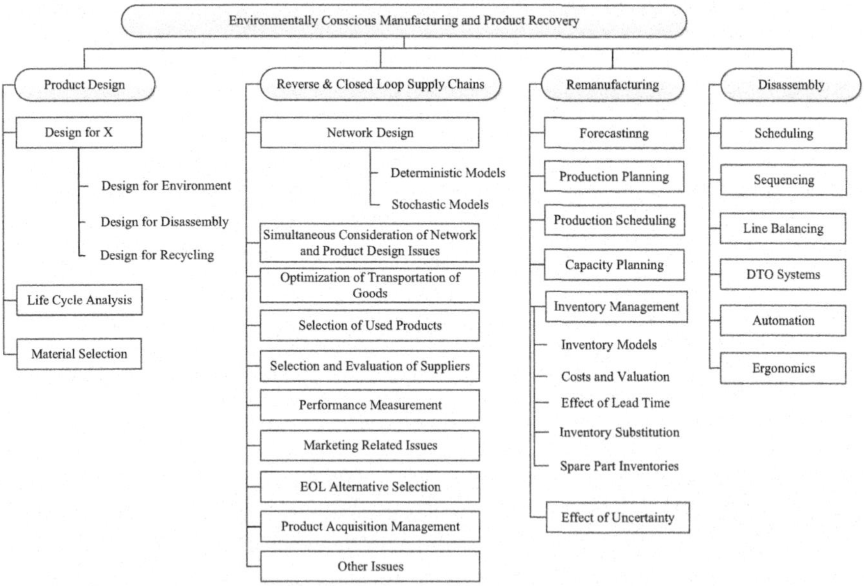

Fig. 1.11 Classification of issues in environmentally conscious manufacturing and product recovery according Ilgin and Gupta (2010)

3.1.2 Lean Manufacturing and the 6R Concept of Sustainable Manufacturing

Sustainable business process management is closely linked to lean production, which was developed by Toyota to continuously improve its production system by focusing on waste reduction and value enhancement in close partnership with its suppliers. Making business processes leaner may result in reducing the environmental impact. Therefore, business process owners could apply the so-called "six R" concept of sustainability, which is the basis of sustainable manufacturing, in their quest to become more sustainable. The 6 R's are: reduce, reuse, recover, redesign, remanufacture, and recycle. This allows for a transformation from an open- loop, single life-cycle paradigm to a theoretically closed-loop, multiple life-cycle paradigm (Jayal et al. 2010) (Fig. 1.12).

In the 6R concept, *reduce* or prevention refers to the reduced use of resources in pre-manufacturing, reduced use of energy and materials, omitting production steps to reduce complexity in the supply chain during manufacturing, and the reduction of waste during the use stage.

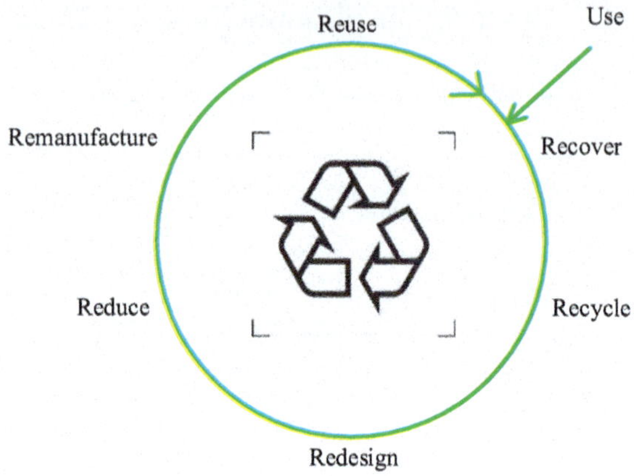

Fig. 1.12 The six R's of sustainable manufacturing according to Jayal et al. (2010)

Reuse refers to the reuse of the product or its components. Reusing or repairing goods can extend their lifespan. This implies using fewer materials and lower energy consumption. Reuse and repair do not only focus on the products in the supply chain, they also include preventive maintenance of equipment to reduce the need for replacement, and selling used equipment that still might be useful for other companies instead of scrapping it.

Recycling encompasses any operation that reprocesses waste materials into products, materials, or substances, whether for their original or other purposes. Recycling originally focused on valuable materials such as metals, but in the last few decades, municipalities and governments started stimulating the separation of goods at the moment of disposal to enhance their recyclability. Examples include the separate collection of plastic bottles, paper, glass, etc.

Recover aims to convert waste into material or energy resources (such as electricity or heat) through thermal and biological means. It should only occur when reduce, reuse, and recycle have failed as preferred options.

Redesigning products aims to make the product more sustainable. It questions the need for types and quantities of products. It involves questions like: "Do we need packaging to ship our semi-fabricated materials?" or "Can we develop products that are recyclable or upgradable at the end of their useful life?" A good example is Fairphone, in the Netherlands, which developed a mobile phone that can be upgraded without throwing away the cell phone.

Remanufacture involves the restoration of used products to their original state or a like-new form through the reuse of as many parts as possible without loss of functionality. Searching for products with a high recycled content or seeking out green products encourages the recuperation of materials, components, and products at the end of their useful life. Making buyers and designers aware of the sustainable opportunities associated with this behavior could foster this type of behavior. It can also be used in marketing to show how sustainable a company is.

Supply chain management is an important business process. In the next section, we will discuss what a sustainability transition from supply chain management to sustainable supply chain management entails.

4 Towards Sustainable Supply Chain Management

4.1 Supply Chain Management

Mujkić et al. (2018) and Chopra and Meindl (2015) state that a supply chain consists of all the interconnected networks and parties involved, directly or indirectly, in fulfilling the demands between suppliers and customers. Manufacturing companies also have their own internal supply chains ranging from product design and product manufacturing to product recovery (see Sect. 3). Basic elements of a supply chain network structure encompass the sourcing of materials and services, transportation at multiple levels, inventory management at multiple levels, and physical storage at multiple levels (see Fig. 1.13). All of these elements impact environmental sustainability through energy consumption and emissions, material usage and reuse, and materials and waste disposal (Kaufmann and Adams 2019).

The flows in a supply chain are not just goods. Chopra and Meindl (2015) distinguish three key flows: (1) products including raw materials, work-in-progress (WIP), sub-assemblies and finished goods; (2) funds including invoices, payments and credits; and (3) information including orders, deliveries, marketing promotions, plant capacities, inventory, etc.

Supply chain management is about optimizing the time and place utility of materials and services. The classic goal of every supply chain is to maximize the supply chain surplus or the overall value generated. The supply chain surplus is defined as the revenue generated from a customer, minus the total cost incurred to produce and deliver the product (Chopra and Meindl 2015).

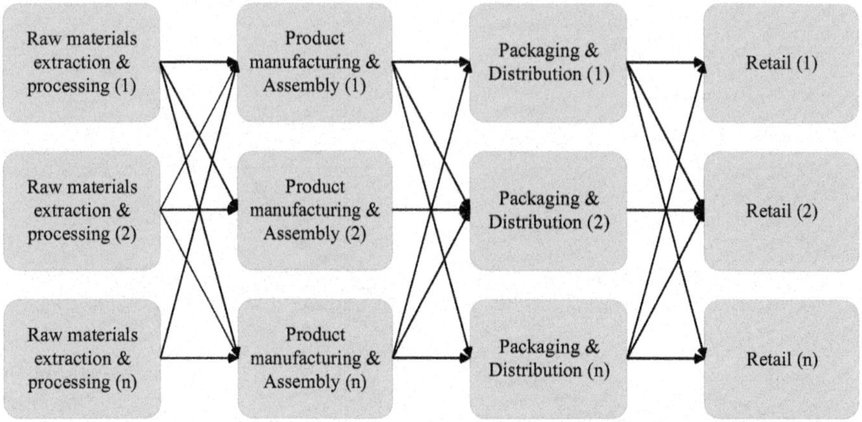

Fig. 1.13 Basic elements of a supply chain network structure

4.2 Sustainable Supply Chain Management

Over the last few decades, supply chain parties focused on their core competencies to become more effective. This resulted in the involvement of an increasing number of parties to satisfy the final customer demand. Consequently, more emphasis has been placed on trust and collaboration amongst supply chain partners to improve their inventory management.

This has also created more traffic between all these parties to realize just-in-time deliveries as part of lean production. Moreover, increased demand for physical products, wider geographical sourcing of supplies, and a broader distribution of finished products have continually increased the demand for freight transport. Recent developments in e-commerce and reverse logistics to realize closed-loop production will further increase the demand for freight transport.

Figure 1.14 shows the world transport evolution since 1975 (OECD 2017). This figure illustrates the close correlation between economic growth, industrial activity, merchandise trade, and maritime cargo transport over the period 1975–2014. The significant increase in the growth of seaborne transport measured relative to global GDP, and industrial activity underscores the development of more complex supply chains in conjunction with containerized shipping on scheduled routes between major world ports. Much of this growth is driven by the emergence of developing economies like Brazil, China, India, and countries in the Middle East.

According to the OECD (2017), CO_2 emissions from transport could increase by 60% by 2050 compared to current levels. If additional measures are not taken, CO_2 emissions from global freight could increase by 160% as international freight

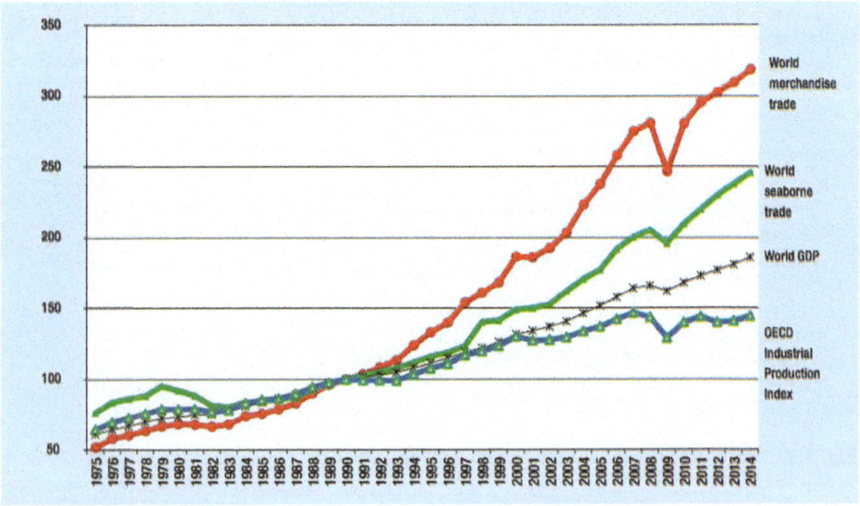

Fig. 1.14 The OECD industrial production index and indices for world GDP, merchandise trade and seaborne shipments (1975–2014) (base year 1990 = 100). (Source: OECD 2017)

volumes are assumed to grow threefold, mainly due to the increased use of road transport. This is especially applicable for short distances and in regions that lack rail links, such as South-East Asia. Optimizing routes or sharing trucks and warehouses between companies could allow higher load factors and fewer empty trips. Such efficiency gains could reduce truck CO_2 emissions by up to one third.

Environmental and societal concerns are absent in the classical supply chain definition of Chopra and Meindl (2015). However, they are gaining importance because of the growing risks they may pose to supply chain stability. Sustainable supply chains taking economic, environmental, and social concerns into account are supposed to be more resilient. The vanilla crisis of 2017 demonstrates how a classic, solely economically optimized supply chain can be impacted by climate change and by relying on production in low-wage countries. In early 2017, the worst tropical cyclone in 13 years hit Madagascar, the world's largest vanilla-producing country. Madagascan wages ($1.50 per day) make it impossible for other countries to compete when prices are below $20 a kilo (Bloomberg 2017). As a result, the supply chain for natural vanilla was very dependent on the production of one country. That same country suffered from environmental problems that may reoccur in the future. This led to excessive prices for vanilla in the supply chains using the natural flavor of this flower (e.g., the production of ice cream) as depicted in Fig. 1.15.

Sustainability in supply chains can be difficult to manage due to the scale and complexity of many supply chains. It involves a transition for suppliers, products,

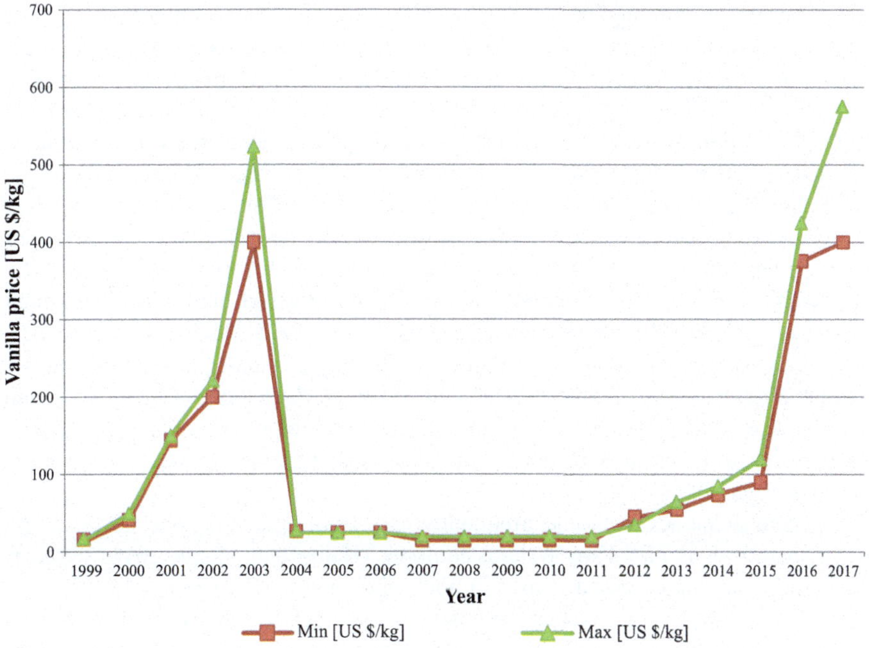

Fig. 1.15 Price increase of vanilla. (Based on https://www.cooksvanilla.com/vanilla-market-report-11-17/)

processes, and customers. Implementing a sustainable supply chain requires considering the type, volume and origin of materials, the design of purchased products, the packaging material and its recyclability, the selected mode of transportation and transportation efficiency, the management of waste and by-products, supplier and customer practices, and investment recovery (Kaufmann and Adams 2019).

Nidumolu et al. (2009) demonstrate that sustainability in business may be a competitive advantage contrary to the belief of many CEOs, particularly in the United States and Europe, that making operations sustainable and developing green products causes a competitive handicap. The benefits of sustainable business operations include efficiency improvements, cost reduction, meeting customer requirements (by complying with branch-specific sustainability requirements such as the Aluminum Stewardship Initiative (ASI) in the aluminum sector), reputation enhancement, environmental protection, and meeting legal and regulatory requirements (e.g., the prohibition of conflict minerals).

However, many barriers to implementing sustainable practices may exist, including a lack of resources for implementation and/or follow up, a lack of knowledge regarding sustainability in general, a lack of knowledge or data concerning supplier practices, and some see no direct connection to improving value.

5 Modeling Sustainable Business Processes

The concept of sustainable development requires different approaches to assess problems related to the economy, ecology, and society. Currently, there is no common view on what sustainability assessment is and how to perform it. This has led to diverse ways of assessing sustainability based on several methodologies (Moldavska and Welo 2016). In general, sustainability issues are characterized by a large number of tradeoffs; thus, compromise solutions need to be discovered (Munda 2005).

Following Munda (2005), we consider Multiple Criteria Decision Analysis (MCDA) to be a suitable tool for assessing sustainability. MCDA offers the appropriate methodological tools to operationalize the concept of incommensurability between economic, environmental, and social objectives at both macro and micro levels of analysis (Munda 2005; Martinez-Alier et al. 1998). MCDA is based on the premise that there are many and, at times, conflicting economic, environmental, and social preferences, but a consensus should be sought (Oglethorpe 2010). According to the stakeholders' viewpoint, the importance of each objective, relative to the three dimensions to be optimized, can be varied by assigning weight factors (Carter and Rogers 2008; Oglethorpe 2010).

System dynamics is another frequently used technique to support sustainability assessments. It is based on systems thinking (Halog and Manik 2011). Systems thinking is an approach that aims to understand emergent behavior in a system by focusing on the whole system and how its interacting parts contribute to that behavior. Dynamic interactions are defined by (1) delays, i.e., differences in time between actions and consequences of these actions; (2) non-linearity, i.e., one action can cause more than one consequence, and one consequence can be caused by more than one action; and (3) feedback loops, both reinforcing and counteracting (Moldavska and Welo 2016).

Following Boulanger and Bréchet (2005), we consider system dynamics to be well suited for supporting sustainable policy development. System dynamics is a powerful modeling technique for explaining and predicting the behavior of real-life supply chains (Towill 1996) by describing the building blocks of the supply chain system and the relationships between them (see Sterman (2000) for more details). System dynamics is also considered to be suitable for modeling the interaction between the earth's resources and human systems (Meadows et al. 1972). However, the application of system dynamics for modeling sustainable manufacturing (Kibira et al. 2009) and sustainable waste management (Morrissey and Brown 2004) has been limited.

Based on the discussion above, Multiple Criteria Decision Analysis and system dynamics are considered to be appropriate tools for assessing sustainable business process management. We will elaborate on the use of Multiple Criteria Analysis in Chaps. 2, 3, 4, 5 and 6, and we will elaborate on using system dynamics in Chaps. 7, 8, 9, 10, 11, and 12.

6 Case Study: Green Waste Valorization

In the following sections, we will gradually explain how to build MCDA models. To do this, we will use a case study entitled "Green Waste Valorization", based on Inghels et al. (2016b, 2019).

Green waste consists of wood cuttings from pruning, leaves, and grass collected after gardening. The cuttings are suitable for composting and energy production since dry wood has an energy content of 18,600 MJ/ton (McKendry 2002a, b). When used as co-firing in a power plant, dry wood can generate, on average, 1650 KWhe/ton. Until recently, green waste could only be used for compost in the EU. The current version of the EU Waste Directive 2008/98/EC (EP&C 2008) permits separating a portion of green waste cuttings for energy recuperation if doing so can be shown to be a more sustainable option. Nevertheless, composting remains the most common option to recover material from green waste (Cesaro et al. 2015).

To better explain the problem setting and the need for a quantitative model to assess the sustainability effects, consider the main options for green waste material/energy recovery depicted in Fig. 1.16. Green waste composted in the open air, also known as *aerobic composting*, only results in compost. It is possible to separate some of the wooden fraction of the green waste and to use it for co-firing in power plants, indicated as *"Pre-treatment"* in Fig. 1.16. When used in Combined Heat Power (CHP) installations, the wooden mass of the green waste can produce both power and heat. The remaining fraction of the green waste can be fermented by means of an *anaerobic digestion* process, which results in biogas that can be added to a natural gas grid after upgrading. The digestate of the anaerobic digestion process can be composted.

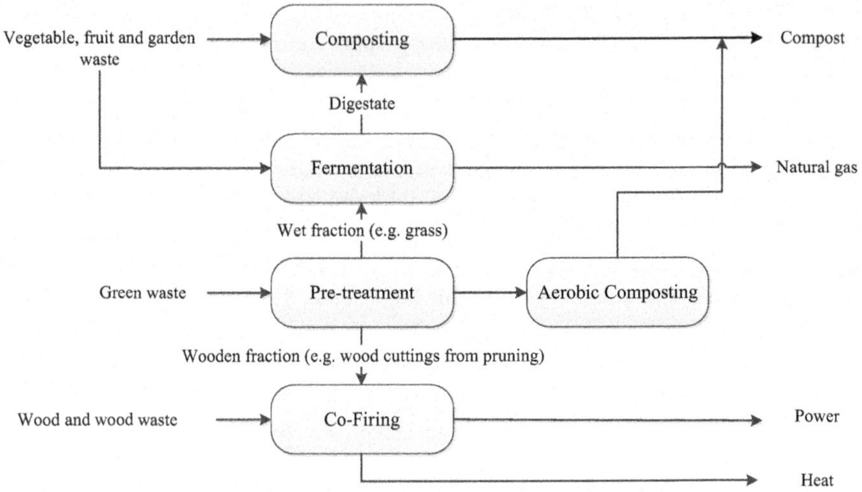

Fig. 1.16 Main options for green waste recovery. (Source: Inghels et al. 2016b)

Anaerobic digestion of green waste as biomass feedstock for renewable energy sources (RES) is not considered to be economically viable (Pick et al. 2012).

A portion of green waste could also be used for energy recuperation as fuel for co-firing in power plants. Doing this reduces CO_2 emissions compared to regular power production (Baxter 2005).

6.1 The Basic System and Its Notation

Figure 1.17 provides an overview of the waste flows in the Green Waste Valorization case and introduces notation based on Inghels et al. (2019). The decision variables in the constrained optimization problem are the mass amounts (in ton) of three different green waste components to be sent to two different destinations, denoted as $x_{c,d}$ (c = component and d = destination). The green waste components $c = \{w,g,l\}$ are wooden prunings "w", leaves "l" and grass "g" (in general, the components w, l and g are referred to as component n in the blend of green waste) and the destinations $d = \{c,i\}$ are composting "c" and incineration "i". The total amount of a component or the total destination is denoted as "*" (e.g., the total amount of wooden material is denoted as $x_{w,*}$ and the total amount of green waste to be composted is denoted as $x_{*,c}$).

The green waste batch will be composted and/or incinerated with energy recovery, which is also known as Waste-To-Energy (WTE). Gases and water are emitted during composting. Gases are also emitted during the WTE process due to the incineration process together with particulate matter (PM). The environmental goal functions Z_j; $j = \{1,2,3,4\}$ are indicated in Fig. 1.17. They are discussed in the following sections.

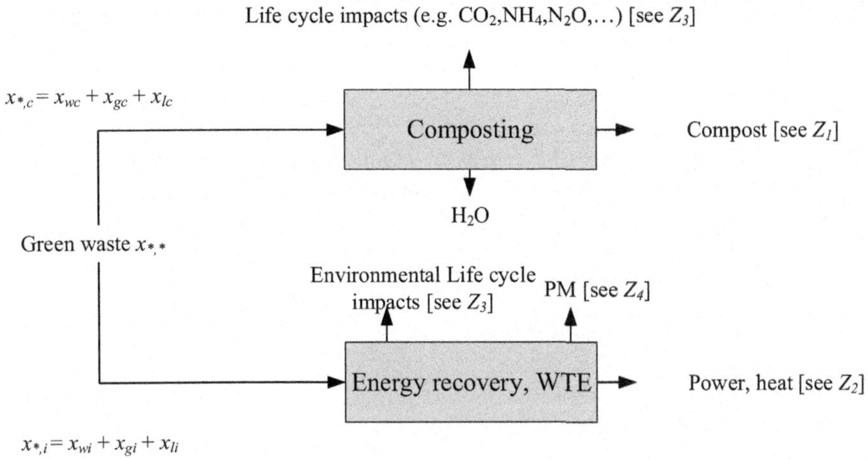

Fig. 1.17 Overview of the waste flows and their environmental impact in the Green Waste Valorization case. (Based on Inghels et al. 2019)

Table 1.1 Characteristics of the main components of green waste

Component n	Carbon content C_n [%wt]	Nitrogen content N_n [%wt]	Dry energy content $LHV_{dry,n}$ [MJ/kg]	Moisture content M_n [%]	Wet energy content $LHV_{wet,n}$ [MJ/kg]
Wood (w)	30.63	0.92	12.60	9.90	11.11
Grass (g)	18.58	0.53	17.04	60	5.35
Leaves (l)	20.23	0.69	11.93	32.95	7.20

Source: Phyllis database; the following numbers indicated with "#" refer to the number in the Phyllis database: Excess fraction wood from organic domestic waste composting plant x_w (#1295), grass x_g (#613), withered leaves x_c (#3065)
Based on Inghels et al. (2019)

We use the Phyllis database (ERCN 2017) to determine the carbon, nitrogen, and moisture content, which are important parameters for the composting process, and the Lower Heating Value (LHV) for determining the energy valorization of green waste as depicted in Table 1.1. The effect of moisture content in biomass on the LHV is expressed in (1.1) (see Fowler et al. 2009). M_n is the moisture content of component n in the blend expressed in %.

$$LHV_{wet,n} = LHV_{dry,n} \cdot (1 - M_n) - 2.442 \cdot M_n \tag{1.1}$$

The total mass of green waste $x_{*,*}$ is assumed to be fully composted and/or incinerated. This is expressed in (1.2)

$$x_{wi} + x_{wc} + x_{gi} + x_{gc} + x_{li} + x_{lc} = x_{*,*} \text{ [ton]} \tag{1.2}$$

6.2 First Environmental Objective: Maximizing Composting Yield

During composting, gasses and water are emitted to the ambient environment resulting in a weight loss expressed in the compost yield factor γ_c that depends on the green waste blend. According to green waste practitioners, an optimal blend of green waste has a compost yield factor of $\gamma_c = 0.35$. Using the compost yield factor, the first environmental objective is maximizing composting yield Z_1:

$$\text{Maximize } Z_1 = x_{*,c} \cdot \gamma_c = (x_{wc} + x_{gc} + x_{lc}) \cdot 0.35[\text{ton}] \tag{1.3}$$

6.3 Second Environmental Objective Function: Maximizing Waste to Energy

The objective function for the energy valorization of green waste by incineration, Z_2, is expressed in (1.4) using lower heating values (LHV) also known as net calorific values (NCV) or lower calorific values (LCV) for wooden prunings, grass, and withered leaves (see Table 1.1). LHV calculations assume that the water component of a combustion process is in a vaporous state at the end of combustion and that the latent heat from the vaporization of water in the fuel and reaction products are not recovered. Note that in Table 1.1, the LHV is expressed in [MJ/kg] and that the mass $x_{n,*}$ of component n is expressed in tons resulting in an energy objective function expressed in [GJ] in (1.4):

$$\text{Maximize } Z_2 = LHV_{wet,wi} \cdot x_{wi} + LHV_{wet,gi} \cdot x_{gi} + LHV_{wet,l} \cdot x_{li}\ [\text{GJ}] \tag{1.4}$$

Using the values of Table 1.1 for the parameters in Eq. (1.4) results in the objective (1.5):

$$\text{Maximize } Z_2 = 11.11 \cdot x_{wi} + 5.35 \cdot x_{gi} + 7.20 \cdot x_{li}[\text{GJ}] \tag{1.5}$$

Note that the functions (1.3) and (1.4) represent opposing trends. Higher volumes of green waste mass assigned to energy recovery, mean lower volumes will be assigned to compost according to Eq. (1.2).

6.4 Third Environmental Objective Function: Minimizing the Environmental Impact

For the environmental impact, we rely on the outcome of the SenterNovem LCA (2008), which represents the total normalized environmental impacts of composting and energy valorization expressed in dimensionless points [Pt]. The life cycle impact categories taken into account are abiotic depletion, global warming, ozone layer depletion, acidification, eutrophication, photochemical oxidant, and eco-toxicity. The total normalized environmental impact of composting and energy recovery is represented by:

$$\text{Minimize } Z_3 = l_c \cdot \left(x_{wc} + w_{gc} + x_{lc}\right) + l_i \cdot \left(x_{wi} + x_{gi} + x_{li}\right) \text{ [Pt]} \quad (1.6)$$

where l_c is the total environmental impact per kg of composted waste and l_i is the total environmental impact per kg of incinerated waste. SenterNovem (2008) gives estimates of these coefficients: $l_c = -575$ pt/kg and $l_w = -1710$ pt/kg. So, Eq. (1.6) becomes:

$$\text{Minimize } Z_3 = -575 \cdot \left(x_{wc} + x_{gc} + x_{lc}\right) - 1710 \cdot \left(x_{wi} + x_{gi} + x_{li}\right) \text{[Pt]} \quad (1.7)$$

The negative sign denotes the net savings in total environmental impact. Although composting and incineration have impacts on the environment, they also offer a saved impact through "avoided" impacts from primary production. Both for composting and incineration, the balance is such that more impact is saved than created.

6.5 Fourth Environmental Objective Function: Minimizing the PM Emissions

Particulate Matter (PM) is only generated and emitted during green waste energy recovery when part of the green waste mixture is incinerated. PM (mostly smaller than 10 microns and easily absorbed by the lungs) induces increased mortality in the event of long-term exposure even at low levels (Annesi-Maesano et al. 2007). Therefore, the emission of PM should be minimized. This can be achieved by installing an air filtration system or by selecting a blend of green waste (wooden prunings, grass, and leaves) that emits a minimum of PM during incineration, respectively denoted as p_w, p_g and p_l. This is expressed in the fourth objective:

$$\text{Minimize } Z_4 = p_w \cdot x_{wi} + p_g \cdot x_{gi} + p_l \cdot x_{li} \text{[kg]} \quad (1.8)$$

For PM coefficients we use $p_w = 21$ kg PM per ton wood (Curtis 2002; EPA 1986), $p_g = 7.08$ kg PM per ton grass (Boubel et al. 1969) and $p_l = 90.20$ kg of PM per ton leaves (Curtis 2002; Battelle 1975; Friedman and Calabrese 1977). This results in (1.9):

$$\text{Minimize } Z_4 = 21 \cdot x_{wi} + 7.08 \cdot x_{gi} + 90.20 \cdot x_{li} \text{ [kg]} \tag{1.9}$$

6.6 Societal Impact

The societal impact will be calculated in chapter six, where we will add additional information to the case. The societal impact is represented by the objective function Z_5 representing the societal impact related to four different ways, $\alpha_{1...4}$, of dealing with green waste. The way with the highest societal impact is the preferred one.

$$\text{Maximize } Z_5 = 0.40 \cdot \alpha_1 + 0.05 \cdot \alpha_2 + 0.23 \cdot \alpha_3 + 0.26 \cdot \alpha_4 \tag{1.10}$$

6.7 Economic Impact

The economic impact, represented by objective function Z_6, is expressed as the profit of composting and incineration of the incoming green waste mass (Table 1.2).

When the options of composting wood, x_{wc}, grass, x_{gc}, and leaves, x_{lc}, are taken into account, together with the incineration of incoming wood, x_{wi}, the profit objective can be formulated as in (1.11). Since the fractions x_{wc}, x_{gc} and x_{lc} represent incoming mass, and the profit of €5/ton is related to the output of the composting

Table 1.2 Overview of cash flow parameters for different investment alternatives

	Investment in additional equipment [€]	Variable cost [€/ton final product]	Revenue [€/ton final product]	Profit [€/ton final product]
Compost	0		5	5
Biomass retrieved from sieve overflow	0	2	6.5	4.5
Biomass (prunings) retrieved from pre-treatment	200,000	2	11	9

Data based on OVAM (2009)

process, the compost yield factor $\gamma_c = 0.35$ is multiplied by the benefit of €5/ton in the first part of the equation, equaling the factor 1.75 in (1.10).

$$\text{Maximize } Z_6 = 1.75 \cdot \left(x_{wc} + x_{gc} + x_{lc}\right) + 9 \cdot x_{wi} \tag{1.11}$$

[€]

6.8 Constraints

The constraints for the Green Waste Valorization case represent limitations to the amount of green waste to be assigned to composting or waste recovery. In general, a conventional blend of green waste consists of a wooden fraction of fresh wooden prunings, $x_{w,*}$, grass, $x_{g,*}$, and leaves, $x_{l,*}$, that can be assigned either to composting or energy recovery. A typical ton of green waste consists of a blend of $\mathrm{x}_{\mathrm{w},*} = 0.6$ ton wooden cuttings, $x_{g,*} = 0.2$ ton grass and $x_{l,*} = 0.2$ ton leaves.

$$x_{wc} + x_{wi} = x_{w,*}[\text{ton}] \tag{1.12}$$

$$x_{gc} + x_{gi} = x_{g,*}[\text{ton}] \tag{1.13}$$

$$x_{lc} + x_{li} = x_{l,*}[\text{ton}] \tag{1.14}$$

We assume that an optimal blend is used for composting. The mass fractions to be composted are calculated using the carbon, nitrogen, and moisture content of the 'regular' compost mixture. This optimal blend can be achieved by applying the following mass balance (Inghels et al. 2019):

$$38.41 \cdot x_{wc} + 47.20 \cdot x_{gc} - 162.45 \cdot x_{lc} = 0[\text{ton}] \tag{1.15}$$

Finally, all masses are expressed in positive numbers.

$$x_{wc}, x_{gc}, x_{lc}, x_{wi}, x_{gi}, x_{li} \geq 0 \tag{1.16}$$

Chapter 2
Single Objective Optimization

In this chapter, we discuss the optimization of sustainability problems with one single objective that is subjected to one or more constraints. This kind of problem is called a *Single Objective Optimization Problem*, abbreviated as *SOOP*. Linear programming (LP) is the mathematical technique we use for solving a SOOP. In linear programming, 'linear' refers to the fact that all the mathematical functions, i.e. the objective function and constraints, are required to be linear. 'Programming' refers to planning problems. In this book, linear programming is used to solve business process planning problems such as the maximization of profit or the minimization of environmental or societal impact.

In the following sections, we will elaborate on how to find optimal solutions for a sustainability SOOP using a *Municipal Solid Waste (MSW) allocation problem*. This problem has two decision variables, which make it possible to find an optimal outcome using three different solution methods: graphically, algebraically, and using the MS Excel Solver function. The solution technique will be applied to the MSW allocation problem as well as to the Green Waste Valorization problem.

1 The Municipal Solid Waste Allocation Problem

We illustrate how to derive a SOOP from a practical business process problem using what we call the *Municipal Solid Waste* (MSW) allocation problem depicted in Fig. 2.1. A municipality collects up to 80 tons of municipal solid waste from households using MSW refuse collection trucks and drops it in an MSW intermediate storage place every day. The MSW consists of goods that are discarded by the public, such as plastic bottles, newspapers, packaging, etc. To dispose of the MSW, the municipality operates two small test setup incinerators, A and B, which incinerate MSW with energy recuperation. This process is also known as Waste-To- Energy or WTE. Both incinerators gather the waste to be processed from the MSW

D. Inghels, *Introduction to Modeling Sustainable Development in Business Processes*, https://doi.org/10.1007/978-3-030-58422-1_2

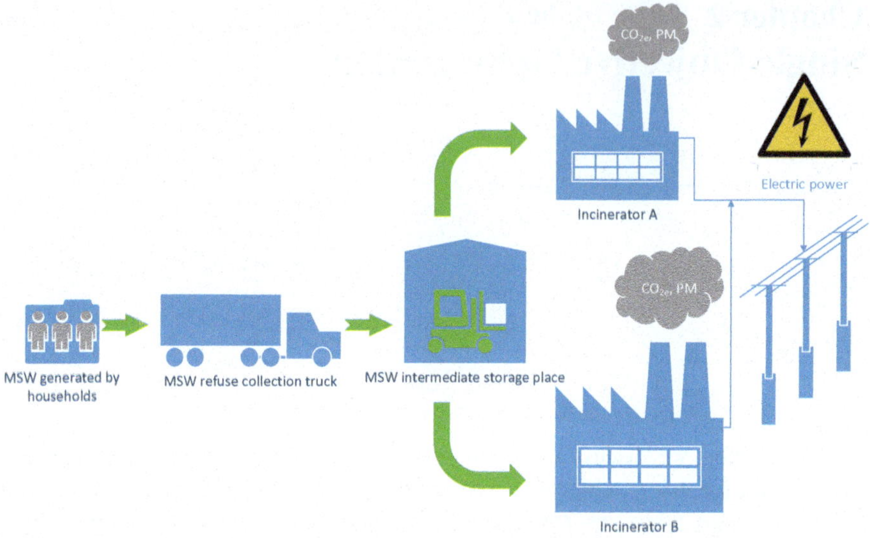

Fig. 2.1 Overview of the MSW allocation problem

intermediate storage place. Incinerator A can process 30 tons of MSW/day and produces power that is sold with an operational profit of €12/ton of MSW. Incinerator B can process 40 tons of MSW/day and produces power that is sold with an operational profit of €15/ton of MSW. The excessive waste that cannot be incinerated is sent to a landfill at zero operational cost. The incineration of MSW generates CO_{2e} emissions (i.e., CO_2 and other greenhouse gases such as CH_4 and N_2O; all are expressed in terms of CO_2 as a life cycle impact). We will further elaborate on life cycle impacts in Chap. 5. Let's assume that Incinerator A emits 0.2 ton CO_{2e} per ton of MSW incinerated, and Incinerator B emits 0.3 ton CO_{2e} per ton of MSW incinerated. The municipality's environmental permit allows for emissions of up to a maximum of 15 tons CO_{2e}/day. The business process planning problem of the MSW Incinerator Manager consists of determining how to generate the maximum operational profit while taking into account the capacity and emission restrictions.

In other words, the question to be solved is to determine:

- the maximum daily operational profit
- the daily volumes that are incinerated in Incinerator A and B under these restrictions. These daily volumes are denoted x_A and x_B and are expressed in [ton/day].

The optimal solution for the MSW allocation problem is obtained using linear programming. This technique is explained theoretically in the next two sections before it is applied to the MSW allocation problem.

2 Standard Form of the Mathematical LP Problem

2.1 *Generic Form of a Linear Programming Problem*

A linear programming problem is any problem that can be formulated as a single objective optimization problem that aims to determine the values for decision variables $x_1, x_2, \ldots, x_n$ so as to

$$\text{Maximize } Z = c_1 \cdot x_1 + c_2 \cdot x_2 + \cdots + c_n \cdot x_n \tag{2.1}$$

Subject to (abbreviated as "s.t.") restrictions

$$a_{11} \cdot x_1 + a_{12} \cdot x_2 + \ldots + a_{1n} \cdot x_n \leq b_1. \tag{2.2}$$

$$a_{21} \cdot x_1 + a_{22} \cdot x_2 + \ldots + a_{2n} \cdot x_n \leq b_2 \tag{2.3}$$

$$\ldots$$

$$a_{m1} \cdot x_1 + a_{m2} \cdot x_2 + \ldots + a_{mn} \cdot x_n \leq b_m \tag{2.4}$$

and

$$x_1, x_2, \ldots, x_n \geq 0 \tag{2.5}$$

The Eq. (2.1) that is maximized is called the *objective function*; it expresses the objective of the problem to be optimized. For sustainability problems, this can include profit maximization, environmental emission minimization, gender equality maximization, etc. The restrictions expressed in (2.2), (2.3) and (2.4) are called *functional constraints.* Constraints express boundary conditions for the decision variables $x_1, x_2, \ldots, x_n$. An example of such a constraint is the production capacity limitation in the MSW planning problem discussed in Sect. 1 of this chapter. The left side of the Eqs. (2.2), (2.3) and (2.4) is commonly referred to as Left Hand Side (LHS) and the right side of thee equations as Right Hand Side (RHS). The constraints expressed in (2.5), which restrict the values of the decision variables, are called *non-negativity constraints.* The constants $c_1, c_2, .. a_{11}, .. a_{mn}, b_1, \ldots b_m$ are called the *parameters* of the model.

Depending on the objective of the problem and the constraints, there are many variations to the standard form described above.

If an objective has to be minimized, such as the minimization of CO_2 emissions, then the objective function will be expressed as:

$$\text{Minimize } Z = c_1 \cdot x_1 + c_2 \cdot x_2 + \cdots + c_n \cdot x_n \tag{2.6}$$

A minimization problem, Minimize $Z_j = f(x_j)$ with decision variables $x_j = x_1, x_2, \ldots, x_n$ can be transformed into a maximization problem by multiplying the objective function by -1 resulting in Maximize $Z'_j = -f(x_j)$. Both objective functions are subjected to the same constraints. Any solution to the maximization problem will also be a solution to the minimization problem and vice versa. (Note that the optimal outcome of the objective function of the maximization problem Z'_j will be -1 times the optimal outcome of the objective function of the minimization problem Z_j. The optimal solution for the decision variables will remain the same).

$$\text{Minimize } Z_j = f(x_j) \text{ is equal to Maximize } Z'_j = -f(x_j) \text{ with } x_j = x_1, x_2, \ldots, x_n \tag{2.7}$$

Constraints might also be expressed as "greater than or equal to" or "equal to" inequalities such as:

$$a_{11} \cdot x_1 + a_{12} \cdot x_2 + \ldots + a_{1n} \cdot x_n \geq b_1 \tag{2.8}$$

$$a_{11} \cdot x_1 + a_{12} \cdot x_2 + \ldots + a_{1n} \cdot x_n = b_1 \tag{2.9}$$

For the constraints a transformation between "greater than or equal to" and "lower than or equal to" inequalities can occur. For example, the constraint (2.8) can be transformed into:

$$-a_{11} \cdot x_1 - a_{12} \cdot x_2 - \ldots - a_{1n} \cdot x_n \leq -b_1 \tag{2.10}$$

Note that in (2.10), the LHS and RHS are both multiplied with "-1" and that the operator is changed from "greater than or equal to" to "lower than or equal to".

Finally, the non-negativity constraints in the standard form (2.5) can be extended with unrestricted decision value constraints or integer value constraints depending on the type of planning problem to be solved.

Any specific value of the decision variables ($x_1, x_2, \ldots, x_n$) is called a *solution* to the problem. Such solutions are called *feasible solutions* if all the constraints are satisfied, and *infeasible solutions* if at least one constraint is violated. A feasible solution can also be an optimal solution. A feasible solution is optimal if its objective function value is equal to the largest value of the objective function (in case of maximization) or the smallest value of the objective function (in case of minimization) over the feasible region.

2.2 *The MSW Allocation Problem Formulated as a Linear Programming Problem*

The MSW allocation problem explained in the introduction of this chapter can be formulated as a linear programming problem as follows:

$$\text{Maximize } Z = 12 \cdot x_A + 15 \cdot x_B [€/\text{day}] \tag{2.11}$$

s.t.

$$0.2 \cdot x_A + 0.3 \cdot x_B \leq 15 [\text{ton/day}] \tag{2.12}$$

$$x_A + x_B \leq 80 \ [\text{ton/day}] \tag{2.13}$$

$$x_A \leq 30 \ [\text{ton/day}] \tag{2.14}$$

$$x_B \leq 40 \ [\text{ton/day}] \tag{2.15}$$

$$x_A, x_B \geq 0 \ [\text{ton/day}] \tag{2.16}$$

Where (2.11) expresses the objective of maximizing the daily operational profit. This operational profit is composed of €12/ton for each ton incinerated in Incinerator A and €15/ton for each ton incinerated in Incinerator B. The total operational profit is the sum of both multiplied by the volumes x_A and x_B incinerated respectively in incinerators A and B. It is evident that if more incoming waste is allocated to incinerator B, this will yield more operational profit since the parameter of the volume incinerated in incinerator B (i.e., €15/ton) is larger than the parameter allocated to incinerator A (i.e., €12/ton). However, the constraints of the problem will also play an important role in the allocation of incoming MSW to incinerator A or B.

The first constraint (2.12) expresses the limitation of CO_2 emissions per day associated with the incineration of the MSW. The CO_2 emissions of Incinerators A and B multiplied with the volume incinerated per incinerator, i.e., x_A and x_B, are added, and this sum must be lower than or equal to the upper bound of 15 ton of CO_2 emissions per day to comply with the maximum allowed emission on the environmental permit.

Next, Eq. (2.13) expresses that the total amount of MSW to be incinerated each day in Incinerator A and B is maximum 80 tons/day and takes into account that there is always sufficient MSW to be incinerated. Eq. (2.13) links the total incoming MSW to both incinerators by splitting it into a fraction x_A allocated to Incinerator A and a fraction x_B to Incinerator B.

Ultimately, the capacity of both incinerators is limited. The capacity constraint of Incinerator A, which is capable of incinerating up to 30 tons/day, is expressed in Eq. (2.14). The capacity constraint of Incinerator B, which is capable of incinerating up to 40 tons/day, is expressed in Eq. (2.15).

Finally, Eq. (2.16) assures that all outcomes of x_A and x_B will be non-negative since physical masses cannot be negative.

The constraints of the MSW allocation problem are depicted in dotted lines in Fig. 2.2.

The constraints (2.12), (2.14), (2.15), and (2.16) limit the zone for feasible decision variables x_A and x_B. These constraints are called "*binding*." Together they form the feasible region. The feasible region is defined as a set of all possible values

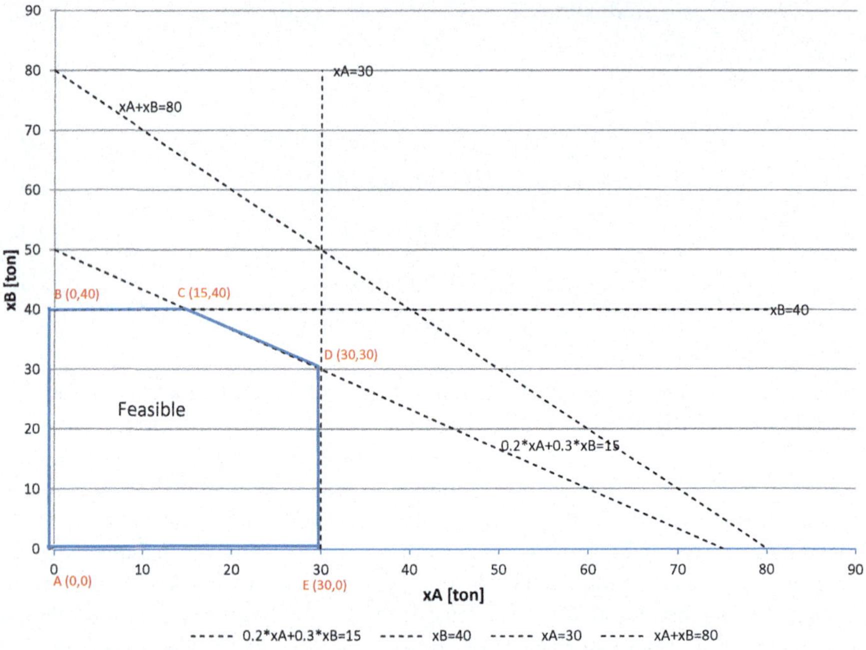

Fig. 2.2 Constraints and feasible region of the MSW allocation problem

of the linear programming problem that satisfies all the constraints of the problem. The feasible set is a polygon. The corner points A, B, C, D and E of the feasible region in Fig. 2.2 are $(x_A, x_B) = (0,0); (0,40); (15,40); (30,30)$ and $(30,0)$ respectively. When these corner points are entered as decision variables into the objective function, the results are called *corner-point feasible solutions* or *CPF solutions.*

Please note that constraint (2.13) will not be of any influence since it does not cross the feasible region and because all values below this line fulfill constraint (2.13). Such a constraint is called "*non-binding*." This constraint could be skipped in solving the SOOP since it will not influence the outcome of the problem.

3 Assumptions of Linear Programming

A problem can be represented as a linear programming problem if the objective function and the constraints are linear in the decision variables and if following assumptions hold (Hillier and Lieberman 2001):

3.1 Proportionality

Proportionality assumption: "*The requirement that the objective function Z and constraints are linear implies the requirement that the value of the objective function Z and the response of each resource expressed by the constraints is proportional to the level of each activity expressed in the variables.*"

This means that the value of the objective function Z is proportional to the level of activity of each of the decision variables x_j as represented by the term $c_j \cdot x_j$ in the objective function. The same applies to the constraints. The contribution of each activity to the left-hand side of each functional constraint is proportional to the level of the activity x_j as represented by the term $a_{ij} \cdot x_j$ in the constraint.

The proportionality assumption rules out that any decision variable x_j can have an exponent other than 1. However, it does not rule out the use of products of two or more decision variables. Therefore, another assumption, i.e., the additivity assumption, is added.

Applied to the MSW allocation problem, the daily operational profit of Incinerator A is proportional to the term $12{\cdot}x_A$. Incinerating 1 ton of MSW in Incinerator A yields €12; incinerating 5 tons of MSW in Incinerator A yields five times €12 or €60.

3.2 Additivity

Additivity assumption: "*The requirement that the objective function Z and constraints are linear implies the requirement that the effects of the value of each variable on the values of the objective function and the constraints are additive. In other words, there can be no interactions between the effects of different activities.*"

This means that the level of activity of decision variable x_i should not affect the level of activity x_j. In other words, the exponent for each decision variable shall be 1, and cross product terms such as $x_i \cdot x_j$ are not allowed amongst decision variables.

3.3 Divisibility

Divisibility assumption: "*decision variables in a linear programming model are allowed to have any values as long as these values satisfy the functional and non-negativity constraints.*"

This means that the decision variables can be integer, binary, non-integer, negative or positive values. Since each decision variable represents the level of some activity, it is assumed that the activities can be run at fractional levels.

3.4 Certainty

Certainty assumption: "*the linear programming model assumes that the responses to the values of the variables are exactly equal to the responses represented by the coefficients.*"

This means that the parameter values c_j, a_{ij} and b_j are assumed to be known constants. For practical problems, this certainty assumption cannot always be satisfied. In many cases, the parameter values are predicted or estimated values, which introduces a degree of uncertainty.

To overcome this problem, a sensitivity analysis is conducted to assess in which range a change of parameter value will not affect the outcome of the linear programming model.

Applied to the MSW allocation problem, we assume that the operational profit of each incinerator is constant and that the daily limit for CO_2 emissions related to the incineration activities is constant. However, it could be possible that the profit erodes because of competition with newly introduced best possible techniques in newly built waste management plants or that regulations for CO_2 emissions become stricter. By conducting a sensitivity analysis, the impact of the possible new parameters on the optimal solution can be tested and evaluated.

3.5 Data Availability

Data availability assumption: "*the formulation of a linear programming model assumes that data are available to specify the associated modeled problem.*"

Data is available for the MSW allocation problem.

4 Determining the Optimal Solution

The optimal solution of a SOOP or linear programming problem can be determined graphically, algebraically, or using the MS Excel Solver. These three possibilities will be described in the next sections using the example of the MSW allocation problem.

4.1 Graphical Method

We first explain the graphical method that can be used to solve any linear programming problem with two decision variables. It is cumbersome to use for three decision variables and impossible to use for problems with more than three decision variables.

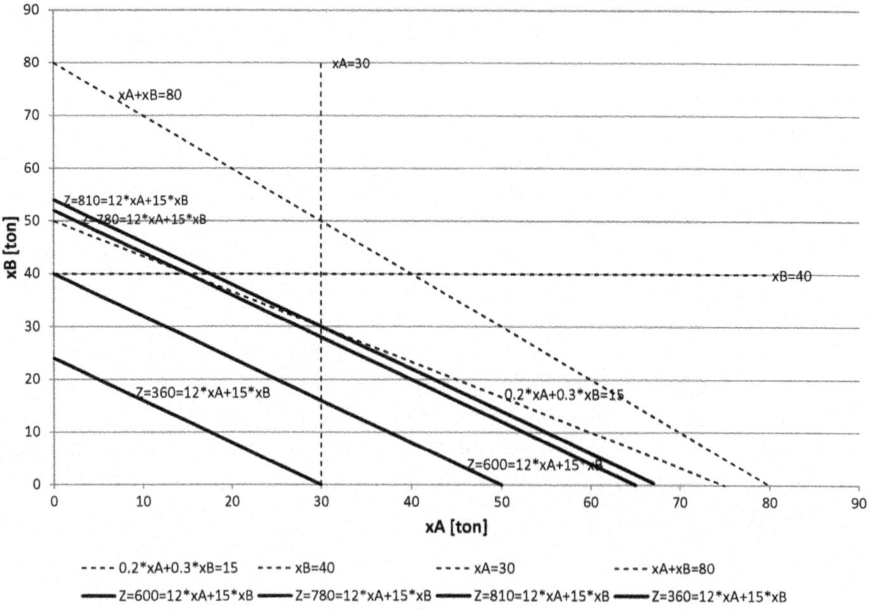

Fig. 2.3 Graphical method to determine the optimal solution for the MSW allocation problem

It can be demonstrated that each optimal solution of a linear programming problem lies at the intersection of a corner point of the feasible region with the objective function. Since the MSW allocation problem is a maximization problem, we examine the objective function Z for values of the decision variables (x_A, x_B) equal to (0,40), (15,40), (30,30) and (30,0). We do not take (0,0) into account. The latter would result in an obvious operational profit of €0/day. The outcome is depicted in Fig. 2.3. It consists of a set of parallel linear solutions with an identical slope intercept:

$$Z = 600 = 12 \cdot x_A + 15 \cdot x_B; (x_A, x_B) = (0, 40)$$

$$Z = 780 = 12 \cdot x_A + 15 \cdot x_B; (x_A, x_B) = (15, 40)$$

$$Z = 810 = 12 \cdot x_A + 15 \cdot x_B; (x_A, x_B) = (30, 30)$$

$$Z = 360 = 12 \cdot x_A + 15 \cdot x_B; (x_A, x_B) = (30, 0)$$

In general, the objective function (2.11) could equivalently be represented by Eq. (2.17) expressing the common slope intercept of the possible solutions to the MSW allocation problem:

$$x_B = -\frac{12}{15} \cdot x_A + \frac{1}{15} \cdot Z \tag{2.17}$$

The maximum operational profit subjected to the constraints is 810 [€/day] for a volume of $x_A = x_B = 30$ [ton/day]. In total Incinerators A and B will incinerate 60 tons of MSW each day. The solution $(x_A, x_B) = (30,30)$ is called the *optimal solution*. There is no better feasible solution that maximizes the operational profit under the given constraint.

The same technique can be applied to a minimization problem. In this case, the solution with the lowest Z-value has to be selected, so (0,0) should not be excluded from the beginning.

The graphical solution is quite straightforward for two decision variables, but this method becomes hard or even impossible for more than two decision variables. This problem can be alleviated using the MS Excel Solver, as we will explain later in this chapter. First, we will introduce the simplex method, which is a commonly used method for solving linear programming problems. This solution method can be selected in the MS Excel Solver.

4.2 Simplex Method

The simplex method, developed by George Dantzig in 1947, is a general procedure for solving linear programming problems that are mostly executed on a computer. This section describes how the MSW allocation problem could be solved using the simplex method.

The simplex method is an algebraic procedure based on underlying geometric concepts. In Fig. 2.2, we determined the corner points of the feasible region. We called the corresponding solutions the corner-point solutions of the feasible solution or corner-point feasible solutions, abbreviated as *CPF solutions*. The corner points in the MSW allocation problem are $(x_A, x_B) = (0,0), (0,40), (15,40), (30,30)$ and $(30,0)$. As Fig. 2.2 depicts, each corner point solution lies on the intersection of two constraint boundaries since it is a two-decision variable problem. This can be generalized for an n decision variable problem where the corner point solutions lie on the intersection of n constraint boundaries. Two adjacent CPF solutions share the same constraint boundaries and are connected by a line segment that is referred to as an *edge* of the feasible region. For example, the CPF solutions (0,40) and (15,40) in Fig. 2.2 are adjacent because they share the same constraint boundary $x_B = 40$.

For every linear programming problem that possesses at least one optimal solution, the following general property applies: "*If a CPF solution has no adjacent CPF solutions that are better, as measured by the objective function Z, then it shall be an optimal solution.*" The simplex method uses this optimality test to determine the optimal solution.

The adjacent CPF solutions for the MSW allocation problem are listed in Table 2.1.

Table 2.1 Adjacent CPF solutions for the MSW allocation problem

CPF solution	The adjacent CPF solutions
A (0,0)	(0,40) and (30,0)
B (0,40)	(0,0) and (15,40)
C (15,40)	(0,40) and (30,30)
D (30,30)	(15,40) and (30,0)
E (30,0)	(0,0) and (30,30)

In the previous section, we figured out graphically that (30,30) is the optimal solution and leads to $Z = 810$ [€/day], which is larger than the corresponding objective values of the two adjacent CPF solutions (15,40) and (30,0) that generate the objective values of $Z = 780$ [€/day] and $Z = 360$ [€/day] respectively.

The simplex method works as follows:

Initialization: Choose CPF solution A (0,0) as the initial CPF solution

Optimality test 1: The two adjacent solutions of A (0,0) with corresponding Z-value $Z_A = 0$, i.e. B (0,40) and E (30,0) have corresponding Z-values $Z_B = 600$ [€/day] and $Z_E = 360$ [€/day] which are both better than Z_A since this is a maximization SOOP.

Iteration 1: Move to a better adjacent CPF solution along the edge that will give a better outcome determined by the coefficients in the goal function. Since the corresponding coefficient of the goal function for x_B, i.e. 15, is larger than for x_A, i.e. 12, it is obvious that moving up the x_B axis will increase the Z-value faster than moving across the x_A axis. At the next CPF solution B (0,40) we execute the optimality test once again:

Optimality test 2: Point C (15,40) leads to a higher Z-value than point B (0,40): $Z_C = 780$ [€/day] > $Z_B = 600$ [€/day].

Iteration 2: Move to the right to point C (15,40) and execute the optimality test again.

Optimality test 3: Point D (30,30) has a higher Z-value than point C (15,40): $Z_D = 810$ [€/day] > $Z_C = 780$[€/day].

Iteration 3: Move to the right, to point D (30,30) and execute the optimality test once again.

Optimality test 4: Point D (30,30) seems to be optimal since the two adjacent CPF solutions C (15,40) and E (30,0) both have a lower Z-value than D: $Z_C = 780$ [€/day] and $Z_E = 360$ [€/day] where $Z_D = 810$[€/day]. Therefore, it can be concluded that Point D is the optimal solution.

4.3 *Algebraic Method*

In this section, we will solve the same MSW allocation problem algebraically using the simplex method. Therefore, the inequality constraints (2.12) – (2.16) first have to be transformed into equality constraints by introducing *slack variables.* Slack variables denote the amount of slack in the left-hand side of the inequality. In the case of the MSW allocation problem, we add the following slack variables: x_1, x_2, x_3 *and* x_4. This leads to what is called the *augmented* form of the MSW allocation problem because adding the slack variables augments the original problem with supplementary variables needed to apply the simplex method. It is also convenient to rewrite the objective function in the same way as the constraints. This leads to the following augmented notation of the MSW allocation problem:

Maximize Z,

Subject to

$$Z - 12x_A - 15x_B = 0 \tag{2.17}$$

$$0.2x_A + 0.3x_B + x_1 = 15 \tag{2.18}$$

$$x_A + x_B + x_2 = 80 \tag{2.19}$$

$$x_A + x_3 = 30 \tag{2.20}$$

$$x_B + x_4 = 40 \tag{2.21}$$

$$\text{and } x_A, x_B, x_1, x_2, x_3, x_4 \geq 0 \tag{2.22}$$

The variables x_A, x_B, x_1, x_2, x_3, x_4 can be grouped in basic and non-basic variables. Since the augmented problem contains more variables than equations, some of the variables must be set to zero to solve the problem. The *basic variables* are defined as the variables which can take any value other than zero. Consequently, the *non-basic variables* are set equal to zero. In case a problem consists of *m* functional constraints and *n* variables, the number of non-basic variables equals *n-m*. The augmented MSW allocation problem contains four functional constraints (2.17) – (2.21) and six variables. Therefore, it has 2 non-basic and 4 basic variables.

It is convenient to write the equations in the tabular form depicted in Table 2.2.

Again, the problem will be solved iteratively. Each iteration consists of three steps. The result is summarized in Table 2.3. For iteration 0, the values from Table 2.2 are copied in Table 2.3.

Table 2.2 Tabular form representation of the MSW allocation problem

Tabular form									
		Coefficient of:							
Eq.	Basic variable	Z	x_A	x_B	x_1	x_2	x_3	x_4	Right side
2.17	Z	1	−12	−15	0	0	0	0	0
2.18	x_1	0	0.2	0.3	1	0	0	0	15
2.19	x_2	0	1	1	0	1	0	0	80
2.20	x_3	0	1	0	0	0	1	0	30
2.21	x_4	0	0	1	0	0	0	1	40

Table 2.3 Simplex tableau for the MSW allocation problem

Iteration	Tabular form											Ratio
	Eq.	Basic Variable	Coefficient of:								RHS	
			Z	x_A	x_B	x_1	x_2	x_3	x_4			
0	2.17	Z	1	-12	-15	0	0	0	0		0	
	2.18	x_1	0	0.2	0.3	1	0	0	0		15	15/0.3=50
	2.19	x_2	0	1	1	0	1	0	0		80	80/1=80
	2.20	x_3	0	1	0	0	0	1	0		30	
	2.21	x_4	0	0	1	0	0	0	1		40	40/1=40
1	2.17	Z	1	-12	0	0	0	0	15		600	
	2.18	x_1	0	0.2	0	1	0	0	-0.3		3	3/0.2=15
	2.19	x_2	0	1	0	0	1	0	-1		40	40/1=40
	2.20	x_3	0	1	0	0	0	1	0		30	30/1=30
	2.21	x_B	0	0	1	0	0	0	1		40	
2	2.17	Z	1	0	0	60	0	0	-3		780	
	2.18	x_A	0	1	0	5	0	0	-3/2		15	
	2.19	x_2	0	0	0	-5	1	0	1/2		25	25/(1/2)=50
	2.20	x_3	0	0	-1	-5	0	1	3/2		15	15/(3/2)=10
	2.21	x_B	0	0	1	0	0	0	1		40	40/1=40
3	2.17	Z	1	0	0	50	0	2	0		810	
	2.18	x_A	0	1	0	0	0	1	0		30	
	2.19	x_2	0	0	0	-10/3	1	-1/3	0		20	
	2.20	x_4	0	0	0	-10/3	0	2/3	1		10	
	2.21	x_B	0	0	1	10/3	0	-2/3	0		30	

Step 1: Determine the entering basic variable, i.e. the variable with the most negative coefficient in row Z, and put a box around the column below this coefficient (see Table 2.3). This column is called the *pivot column* (depicted as a black rectangular box in the simplex tableau).

Step 2: Determine the leaving basic variable by applying the minimum ratio test as follows:

- Select each coefficient in the pivot column that is greater than 0
- Divide the right-side column value for the same row by each of these coefficients
- Identify the row that has the smallest ratio (the dark grey colored row) as the pivot row
- The basic variable in the pivot row in the pivot column is the leaving basic variable

Step 3: The new basic feasible solution is calculated using elementary row operations to construct a new simplex tableau for the next iteration as follows:

- Divide each coefficient in the pivot row (the dark grey row in Table 2.3 for each iteration step) by the leaving basic variable and use this new pivot row in the next steps. The leaving basic variable becomes '1' in the new pivot row.
- For each other row that has a negative coefficient in the pivot column, including the row with the objective function Z, add to each coefficient of this row the product of the absolute value of this coefficient and the corresponding coefficient of the new pivot row.
- For each other row that has a positive coefficient in the pivot column, subtract from each coefficient in this row the product of this coefficient and the coefficient of the new pivot row.
- The results from step 3 are noted in the box for the next iteration. All coefficients in the pivot column become '0' except for the leaving basic variable, which remains '1'.

Applied to the MSW allocation problem example:

Iteration 0: The entering basic variable is x_B, x_4 is the leaving basic variable; row x_4 is the pivot row and column x_B is the pivot column.
Iteration 1: The entering basic variable is x_A, x_1 is the leaving basic variable; row x_1 is the pivot row and column x_A is the pivot column
Iteration 2: The entering basic variable is x_4, x_3 is the leaving basic variable; row x_3 is the pivot row and column x_4 is the pivot column.
Iteration 3: Since there is no negative number left in the Z row, the optimal value is $Z = 810$

We work out iteration 0 as an example. The other iterations are similar:

Step 1: The entering basic value in row Z is the coefficient of x_B, i.e., -15, because this is most negative coefficient in row Z. The column linked to variable x_B is the pivot column.
Step 2: The leaving basic value is the coefficient of x_4, i.e., 1, because this element in the pivot column has the lowest ration, i.e., 40.
Step 3: the new simplex tableau for the next iteration, i.e., iteration 1, is now calculated. Since the leaving basic variable is '1', the pivot row remains unchanged when dividing each coefficient in the pivot row by the leaving variable. Next, we multiply each coefficient in row x_4 with $|-15| = 15$, and each coefficient of row x_4 is added with the corresponding coefficient of row Z. The new coefficients of each row are added to form the new coefficient for row Z in the next iteration 1 (Table 2.4).

Each other coefficient in iteration 0 in the pivot column is positive. The calculation for each of the rows in iteration 1 is similar to the example for row x_1 depicted below (Table 2.5):

Table 2.4 Coefficient calculation for row Z (iteration 1)

Basic Variable	Coefficient of:							Right side
	Z	x_A	x_B	x_1	x_2	x_3	x_4	
Z (iteration 0)	1	-12	-15	0	0	0	0	0
x_4 (iteration 0)	+0*15	+0*15	+1*15	+0*15	+0*15	+0*15	+1*15	+40*15
Z (iteration 1)	= 1	= -12	= 0	= 0	= 0	= 0	= 15	= 600

Table 2.5 Coefficient calculation for row x_1 (iteration 1)

Basic Variable	Coefficient of:							Right side
	Z	x_A	x_B	x_1	x_2	x_3	x_4	
x_1 (iteration 0)	0	0.2	0.3	1	0	0	0	15
x_4 (iteration 0)	-0*0.3	-0*0.3	-1*0.3	-0*0.3	-0*0.3	-0*0.3	-1*0.3	-40*0.3
x_1 (iteration 1)	= 0	= 0.2	= 0	= 1	= 0	= 0	= -0.3	3

The same calculations are repeated for each iteration until all coefficients of the decision variables in row Z are 0, and the coefficient of the Z coefficient is 1. The RHS value for the Z-value is than the optimal solution.

4.4 *Using the MS Excel Solver*

In this section, we will use the MSW allocation problem once again as an example to demonstrate how to solve a SOOP or linear programming problem using the Microsoft (MS) Excel Solver. The MS Excel Solver uses the simplex method to find the optimal solution(s) for the SOOP. The step-by-step approach applied to the MSW allocation problem can be used to solve any other SOOP.

4.4.1 Step 1: Arranging the SOOP Problem

The MS Excel Solver needs the following inputs in a structured way: objective function outcome (in one cell), the decision variables (in a range), the constraints (with LHS, RHS cells input separately together with the ≤, ≥, or = operator signs). We use a commonly used approach to structure the underlying linear programming problem of the SOOP in an Excel table format as follows (for a more extensive discussion see Hillier and Lieberman 2001) (Table 2.6):

Table 2.6 Generic presentation of a SOOP in Excel based on Hillier and Lieberman (2001)

Eq.		x_1	x_2	...	x_n	Total		Available
2.2	Constraint 1	a_{11}	a_{12}		a_{1n}	$= a_{11}*x_1+ ...+ a_{1n}*x_{1n}$	≤	b_1
2.3	Constraint 2	a_{21}	a_{22}		a_{2n}	$= a_{21}*x_1+ ... +a_{2n}*x_{2n}$	≤	b_2
	...							
2.4	Constraint n	a_{m1}	a_{m2}	...	a_{mn}	$= a_{m1}*x_1+ ...+ a_{mn}*x_{mn}$	≤	b_m
2.1	Goal function	c_1	c_2	...	c_n	$Z=c_1*x_1+...+c_n*x_n$		
	solution	x_1	x_2	...	x_n			

Table 2.7 The MSW allocation presentation in Excel

Eq.		x_A	x_B	Total		Available	Dimension
2.12	CO_2 emissions	0.2	0.3	$0.2*x_A+0.3*x_B$	≤	15	[ton/day]
2.13	Total capacity	1	1	x_A+x_B	≥	80	[ton/day]
2.14	Capacity Incinerator A	1	0	x_A	≤	30	[ton/day]
2.15	Capacity Incinerator B	0	1	x_B	≤	40	[ton/day]
2.11	Unit operational profit	12	15	$12*x_A+15*x_B$			[€/day]
	Solution	x_A	x_B				

This structure will help to couple the model parameters and variables to the MS Excel Solver in the next step.

Applied to the MSW allocation problem, the Excel table looks like (see Table 2.7):

The parameters of the model belonging to the decision variables x_A and x_B are listed separately for the constraints (2.12) – (2.15) and for the objective function (2.11). The grey cells in Table 2.7, will be filled in once the Solver has calculated the optimal solution.

4.4.2 Step 2: Linking the Solver Table to the Problem

If the MS Excel Solver is not installed, one must first install it as an "add in". Once the MS Excel Solver is successfully installed on the PC, the Solver function will be displayed on the *Data* tab as depicted in Fig. 2.4.

First, the linear programming problem, as depicted in Table 2.7, must be entered in the spreadsheet as shown in Fig. 2.5. Column B denotes the names of the constraints (rows 5–8), the objective function (row 9), and the value of the decision variables that will contain the values associated with the solution of the linear programming problem after executing the MS Excel Solver (row 10). Columns C and D contain the parameters for the decision variables x_A and x_B respectively. In column E, the LHS algebraical expression of the constraint and the objective

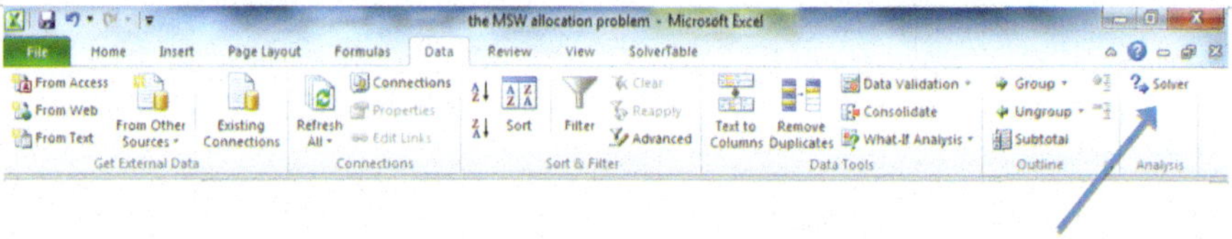

Fig. 2.4 Displaying of the installed Solver in Excel

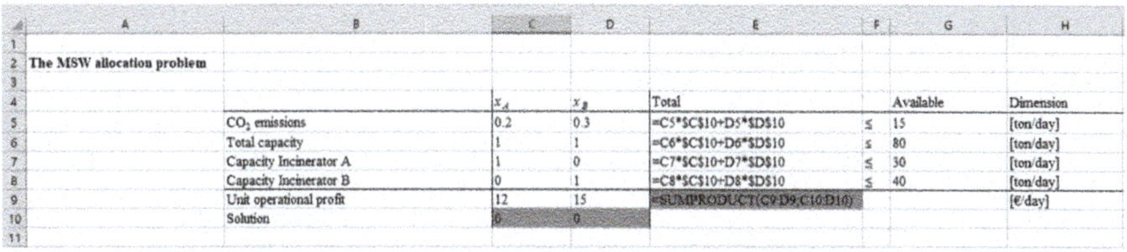

	A	B	C	D	E	F	G	H
1								
2	**The MSW allocation problem**							
3								
4			x_A	x_B	Total		Available	Dimension
5		CO_2 emissions	0.2	0.3	=C5*$C$10+D5*$D$10	≤	15	[ton/day]
6		Total capacity	1	1	=C6*$C$10+D6*$D$10	≤	80	[ton/day]
7		Capacity Incinerator A	1	0	=C7*$C$10+D7*$D$10	≤	30	[ton/day]
8		Capacity Incinerator B	0	1	=C8*$C$10+D8*$D$10	≤	40	[ton/day]
9		Unit operational profit	12	15	=SUMPRODUCT(C9:D9,C10:D10)			[€/day]
10		Solution	0	0				
11								

Fig. 2.5 Entering the MSW allocation problem in Excel

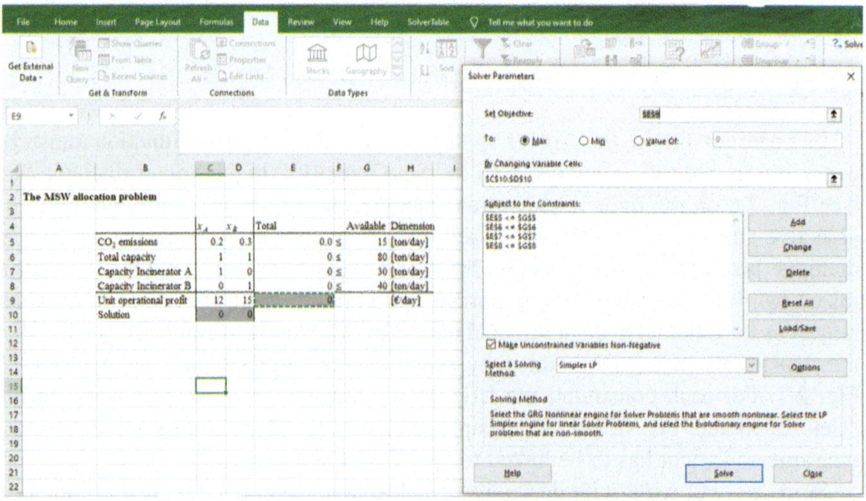

Fig. 2.6 Configuring the Solver Parameters for the MSW allocation problem

functions must be entered. This can be done in two ways: (i) the straightforward way whereby each parameter for every single expression is linked to its corresponding decision variables. This is the case for the expressions in the cells E5 up to E8 and (ii) the condensed way. Instead of writing the expression in full, as in cells E5–E8, MS Excel offers the function SUMPRODUCT that will sum the product of each of the individual terms in the ranges of the corresponding cells. For example, the expression in cell E9 = C9*$C$10 + D59$D$10 can also be written as E9 = SUMPRODUCT(C9:D9,C10:D10).

Once the linear programming problem is arranged correctly, the cells can be linked to the proper input fields in the Solver dialogue box as depicted in Fig. 2.6.

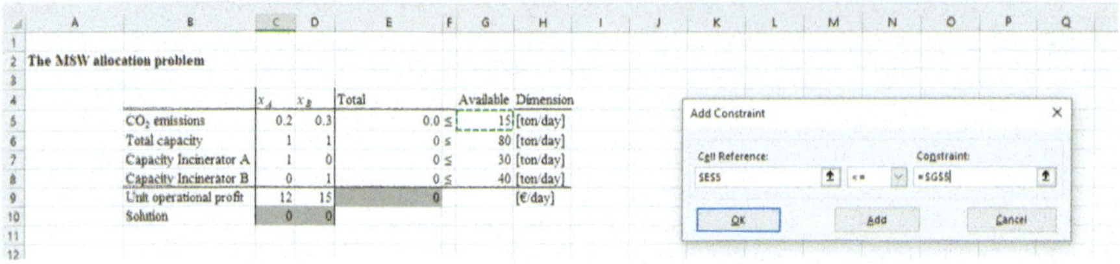

Fig. 2.7 Adding a constraint in Excel Solver

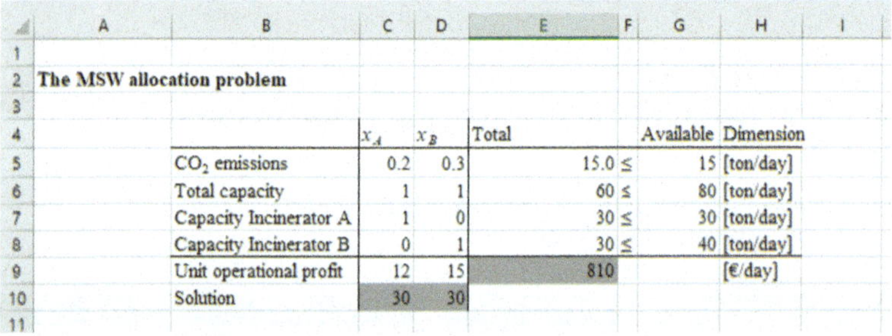

Fig. 2.8 Display of the solution of the MSW allocation problem in Excel

In order to fill in the Solver dialogue box, the objective function must be expressed in one cell and added next to "Set Objective" in the Solver dialogue box (see Fig. 2.6 on the right side). In this example, the MSW allocation problem, the objective function is expressed in cell E9. Then one must select in the Solver dialogue box whether the objective function should be maximized (Max) or minimized (Min) by selecting the appropriate radio button. Next, the decision variables range has to be linked in the input field under "By Changing Variable Cells." Finally, the constraints have to be added by selecting the "Add" button, as depicted in Fig. 2.7. For each constraint, the LHS of the constraint has to be entered in the "Cell Reference" box (E5 in the example of Fig. 2.7), and the corresponding RHS of the same constraint has to be linked to the "Constraint" box.

Since we will work with decision variables larger than or equal to zero, the box "Make Unconstrained Variables Non-negative" should be ticked (see Fig. 2.6) in the Solver dialogue box.

Once this is all done, select "Simplex LP" as the solving method below the Solver dialogue box.

4.4.3 Solving the LP Problem

The problem can now be solved. The Excel table will be filled with the optimal data for the objective function and the corresponding optimal values for the decision variables as outputs, as depicted in dark grey in Fig. 2.8.

4.4.4 Analyzing the SOOP

Next, it is possible to explore the results by downloading the Answer, Sensitivity, and Limits reports that become available in the Solver Results dialogue box (see Fig. 2.9).

The first report is the *Answer Report* (see Fig. 2.10 for the MSW allocation problem). It contains the original values for the constraint and the objective function before the execution of the MS Excel Solver and the optimal result for the objective function and the corresponding values for the decision variables after the execution of the MS Excel Solver. Furthermore, the binding and non-binding constraints are indicated together with the slack value for each constraint, i.e., the difference per constraint between its RHS value and the value for the LHS of the constraint using the optimal values for the decision variables.

The next available report is the *Sensitivity Report* (see Fig. 2.11 for the MSW allocation problem). It provides classical sensitivity analysis information for the linear programming problem or SOOP: i.e., dual values that are named "Reduced Cost" for the decision variables and "Shadow Price" for the binding constraints and range information.

The *Limits Report* (see Fig. 2.12 for the MSW allocation problem) provides some "sensitivity analysis" information. It is created by re-running the Solver model and minimizing and maximizing the objective while keeping all other variables fixed. In return, it shows a "Lower Limit" for each variable, which is the outcome of the minimization problem for the same objective function and constraints. It delivers the smallest value for each decision while satisfying the constraints and holding all of the other variables constant, denoted as "Lower Limit", and the largest value the variable can take under the same restrictions, denoted as "Upper Limit."

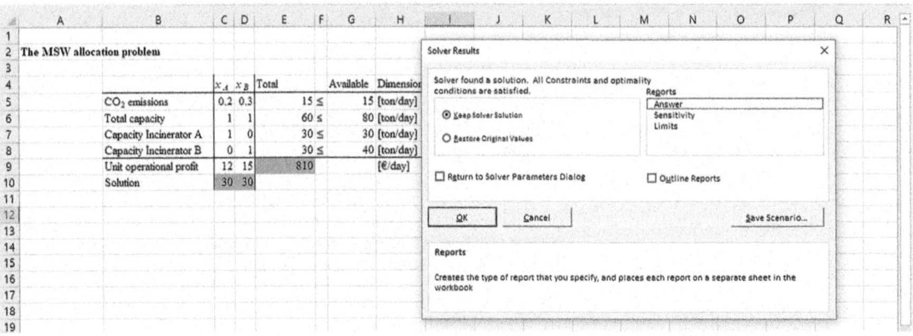

Fig. 2.9 Exploration possibilities of the solved MSW allocation problem in Excel

Microsoft Excel 14.0 Answer Report
Worksheet: [the MSW allocation problem.xlsx]MSW allocation
Report Created: 14-8-2017 20:32:05
Result: Solver found a solution. All Constraints and optimality conditions are satisfied.
Solver Engine
Solver Options

Objective Cell (Max)

Cell	Name	Original Value	Final Value
$E$9	Unit operational profit Tot	0	810

Variable Cells

Cell	Name	Original Value	Final Value	Integer
$C$1	Solution xA	0	30	Contin
$D$1	Solution xB	0	30	Contin

Constraints

Cell	Name	Cell Value	Formula	Status	Slack
$E$5	CO2 emissions Total	150	$E$5<=$G$5	Binding	0
$E$6	Total capacity Total	60	$E$6<=$G$6	Not Binding	20
$E$7	Capacity Incinerator A Tota	30	$E$7<=$G$7	Binding	0
$E$8	Capacity Incinerator B Tota	30	$E$8<=$G$8	Not Binding	10

Answer Report 1 | Sensitivity Report 1 | Limits Report 1

Fig. 2.10 The Answer report for the MSW allocation problem

Microsoft Excel 14.0 Sensitivity Report
Worksheet: [the MSW allocation problem.xlsx]MSW allocation
Report Created: 14-8-2017 20:32:06

Variable Cells

Cell	Name	Final Value	Reduced Cost	Objective Coefficient	Allowable Increase	Allowable Decrease
$C$10	Solution xA	30	0	12	1E+30	2
$D$10	Solution xB	30	0	15	3	15

Constraints

Cell	Name	Final Value	Shadow Price	Constraint R.H. Side	Allowable Increase	Allowable Decrease
$E$5	CO2 emissions Total	150	5	150	30	90
$E$6	Total capacity Total	60	0	80	1E+30	20
$E$7	Capacity Incinerator A Total	30	2	30	45	15
$E$8	Capacity Incinerator B Total	30	0	40	1E+30	10

Answer Report 1 | Sensitivity Report 1 | Limits Report 1 | Answer Report 2 | Sensitivity R...

Fig. 2.11 The Sensitivity report for the MSW allocation problem

Microsoft Excel 14.0 Limits Report
Worksheet: [the MSW allocation problem.xlsx]MSW allocation
Report Created: 14-8-2017 20:32:06

Cell	Objective Name	Value
$E$9	Unit operational profit Total	810

Cell	Variable Name	Value	Lower Limit	Objective Result	Upper Limit	Objective Result
$C$10	Solution xA	30	0	450	30	810
$D$10	Solution xB	30	0	360	30	810

Answer Report 1 | Sensitivity Report 1 | Limits Report 1 | Answer Report 2

Fig. 2.12 The Limits report for the MSW allocation problem

5 Special Cases

In general, a SOOP has one optimal solution. However, there are some exceptions to this rule. In some special cases, multiple optimal solutions can be the outcome when the objective function falls together with the constraint containing the optimal solution values. In the MSW allocation problem, this would be the case if the goal function became, e.g., Maximize $Z = 10 \cdot x_A + 15 \cdot x_B$. In this case, the slope of the objective function falls together with the slope of the constraint (2.12), i.e., $0.2 \cdot x_A + 0.3 \cdot x_B \leq 15$ [ton/day]. Since constraint (2.12) holds the optimal solutions between $(x_A, x_B) = (15{,}40)$ and $(x_A, x_B) = (30{,}30)$, there are a set of equally optimal solutions Z = 750 [€/day] situated between these two points on the bold line in Fig. 2.13.

It is also possible that no optimal solution exists. This would be the case if an extra constraint such as $2 \cdot x_A + 3 \cdot x_B \geq 250$ was added to the existing constraints, as depicted in Fig. 2.14. The arrows on the constraints in Fig. 2.14 indicate the direction in which the feasible solutions will be situated starting from the constraint lines. Obviously, there is no feasible region in the case of this extra constraint since no polygon with feasible corner solution points can be made.

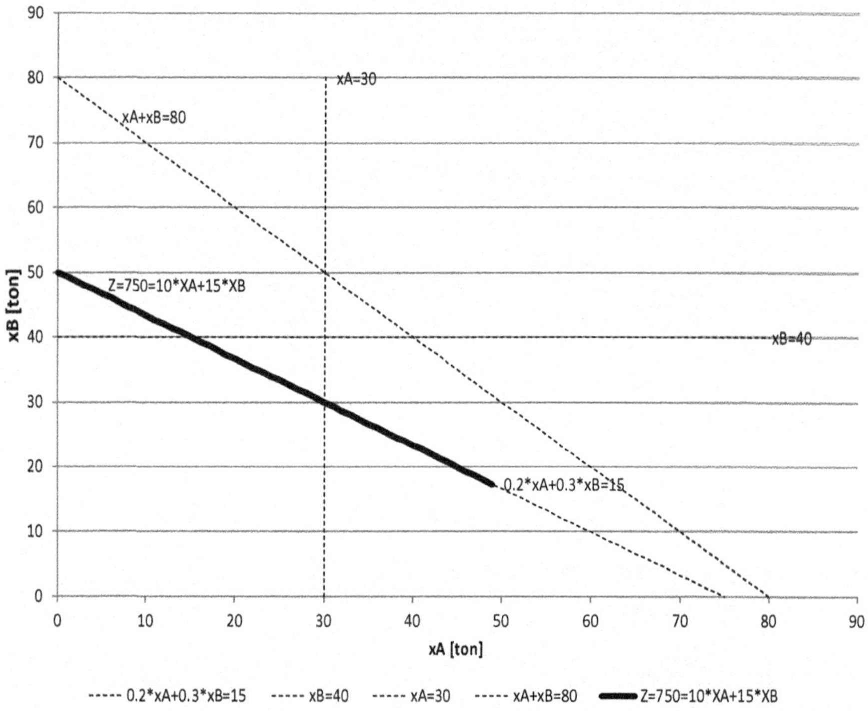

Fig. 2.13 Objective function falls together with a constraint that holds the optimal values of the linear programming problem

6 Optimizing Profit for the Green Waste Valorization Case

Until now, we have only discussed a problem with two decision variables for the convenience of solving the same SOOP using different methods. Real-life problems often have more than two variables, which makes it cumbersome or even impossible to solve them graphically. Using the MS Excel Solver is an easy way to solve SOOPs with many decision variables. This also applies to the Green Waste Management problem introduced in Chap. 1 since this problem has six decision variables. If we only maximize profit, expressed in Z_6 (see (1.11) in Chap. 1), for this problem, the SOOP can be expressed as the following LP problem for a blend of 1 ton of incoming green waste consisting of $x_{w,*} = 0.6$ ton of wooden prunings, $x_{g,*} = 0.2$ ton of grass and $x_{l,*} = 0.2$ ton of leaves:

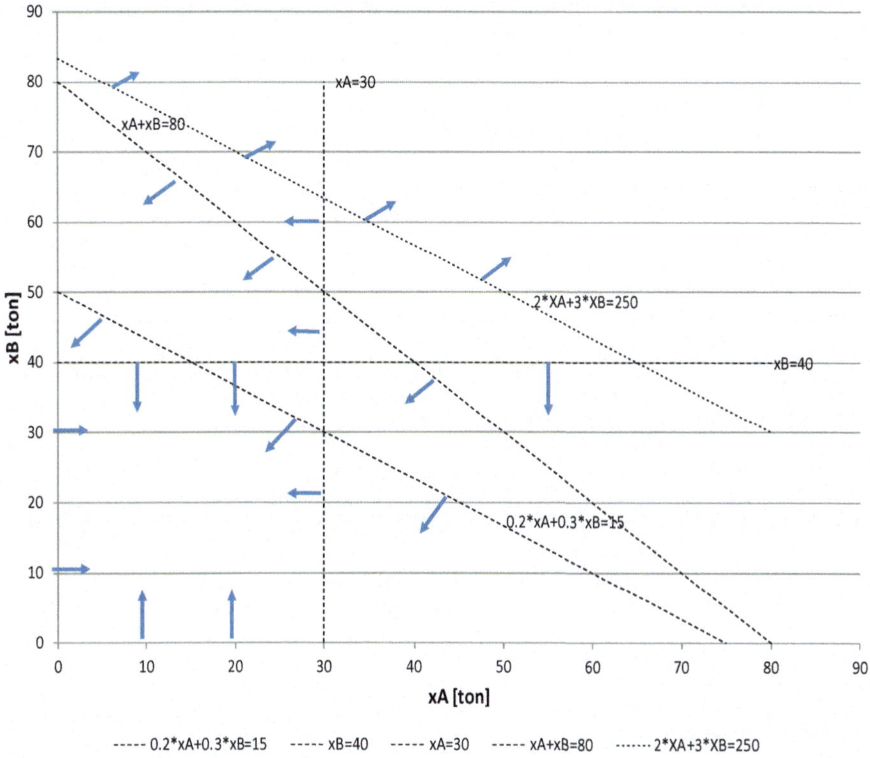

Fig. 2.14 No feasible solution

$$\text{Maximize } Z_6 = 1.75 \cdot (x_{wc} + x_{gc} + x_{lc}) + 9 \cdot x_{wi} \text{ [€]} \tag{1.11}$$

s.t.:

$$x_{wc} + x_{wi} = 0.6 \text{ [ton]} \tag{2.23}$$

$$x_{gc} + x_{gi} = 0.2 \text{ [ton]} \tag{2.24}$$

$$x_{lc} + x_{li} = 0.2 \text{ [ton]} \tag{2.25}$$

$$38.41 \cdot x_{wc} + 47.20 \cdot x_{gc} - 162.45 \cdot x_{lc} \text{[ton]} \tag{1.15}$$

$$x_{wc}, x_{gc}, x_{lc}, x_{wi}, x_{gi}, x_{li} \geq 0 \tag{1.16}$$

The constraints (2.23) – (2.25) are the same as discussed in (1.12) – (1.14) applied to one ton of incoming green waste.

The Excel table for this problem is expressed in Fig. 2.15 using the generic presentation of Table 2.6. The decision variables are the amounts of green waste fractions (wooden prunings, grass, and leaves) allocated to composting or incineration ($x_{wc}, x_{gc}, x_{lc}, x_{wi}, x_{gi}, x_{li}$). The functions expressed in column I correspond with the expressions denoted in column A.

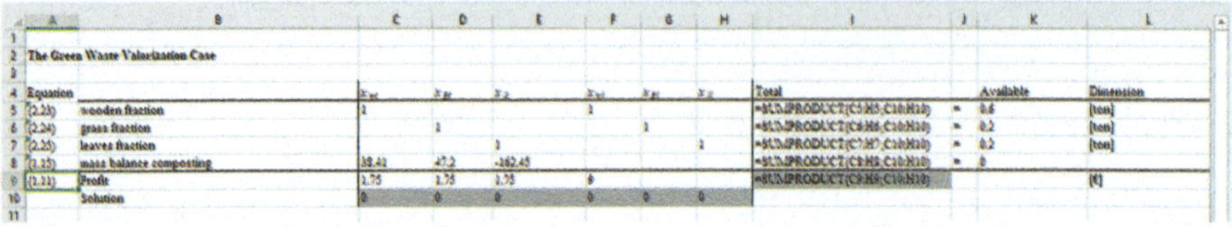

The Green Waste Valorization Case											
Equation		x_{wc}	x_{gc}	x_{lc}	x_{wi}	x_{gi}	x_{li}	Total		Available	Dimension
(2.23)	wooden fraction	1			1			=SUMPRODUCT(C5:H5,C10:H10)	=	0.6	[ton]
(2.24)	grass fraction		1			1		=SUMPRODUCT(C6:H6,C10:H10)	=	0.2	[ton]
(2.25)	leaves fraction			1			1	=SUMPRODUCT(C7:H7,C10:H10)	=	0.2	[ton]
(1.15)	mass balance composting	38.41	47.2	-162.45				=SUMPRODUCT(C8:H8,C10:H10)	=	0	
(1.11)	Profit	1.75	1.75	1.75	9			=SUMPRODUCT(C9:H9,C10:H10)			[€]
	Solution	0	0	0	0	0	0				

Fig. 2.15 SOOP for profit maximization for the Green Waste Valorization case displayed in Excel

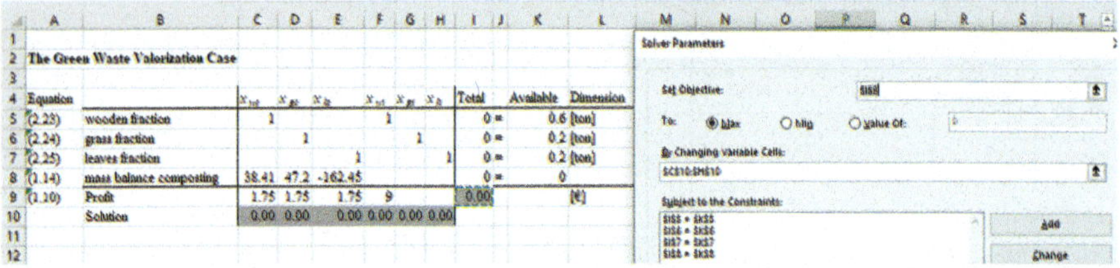

The Green Waste Valorization Case											
Equation		x_{wc}	x_{gc}	x_{lc}	x_{wi}	x_{gi}	x_{li}	Total		Available	Dimension
(2.23)	wooden fraction	1			1			0	=	0.6	[ton]
(2.24)	grass fraction		1			1		0	=	0.2	[ton]
(2.25)	leaves fraction			1			1	0	=	0.2	[ton]
(1.14)	mass balance composting	38.41	47.2	-162.45				0	=	0	
(1.10)	Profit	1.75	1.75	1.75	9			0.00			[€]
	Solution	0.00	0.00	0.00	0.00	0.00	0.00				

Fig. 2.16 The Solver dialogue box for profit maximization of the Green Waste Valorization case

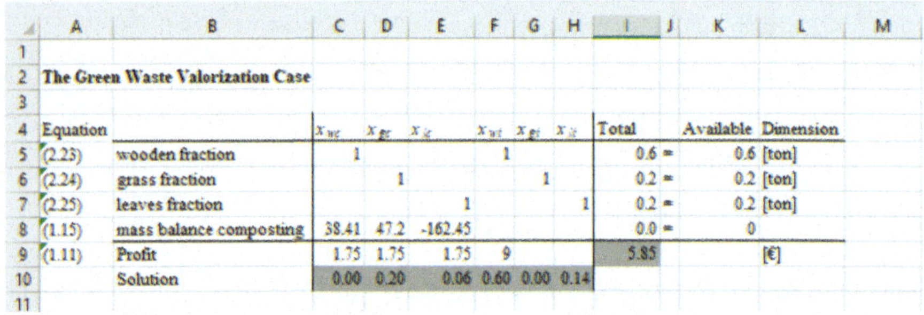

The Green Waste Valorization Case											
Equation		x_{wc}	x_{gc}	x_{lc}	x_{wi}	x_{gi}	x_{li}	Total		Available	Dimension
(2.23)	wooden fraction	1			1			0.6	=	0.6	[ton]
(2.24)	grass fraction		1			1		0.2	=	0.2	[ton]
(2.25)	leaves fraction			1			1	0.2	=	0.2	[ton]
(1.15)	mass balance composting	38.41	47.2	-162.45				0.0	=	0	
(1.11)	Profit	1.75	1.75	1.75	9			5.85			[€]
	Solution	0.00	0.20	0.06	0.60	0.00	0.14				

Fig. 2.17 The optimal solution for profit maximization of the Green Waste Valorization case of one ton of green waste

The MS Excel Solver dialogue box for this problem is depicted in Fig. 2.16. The objective function that is to be maximized is expressed in cell I9. The RHS and LHS parts of the constraints (2.23) – (2.25) and (1.15) are inserted in the constraints input field. The non-negativity constraints (1.16) are guaranteed by selecting the box "Make Unconstrained Variables Non-Negative".

The optimal solution for this SOOP results in the allocation of one ton of green waste to composting with a grass fraction of $x_{gc} = 0.20$ ton and a leaves fraction of $x_{lc} = 0.06$ ton, and to incineration with a wooden prunings fraction of $x_{wi} = 0.60$ ton and a leaves fraction of $x_{li} = 0.14$ ton. In other words, the entire grass fraction in the green waste (0.20 ton) will be composted, and the entire wooden prunings fraction (0.60 ton) will be incinerated, whereas the leaves fraction will be composted (0.06 ton) and incinerated (0.14 ton). The profit will be 5.85 [€/ton]. The results are expressed in Fig. 2.17.

7 Mixed Integer Linear Programming (MILP)

In this last section, we briefly explain the concept of Mixed-Integer Linear Programming (MILP) since it can be considered an extension of Linear Programming. A linear programming problem with some "regular" (continuous) decision variables and some variables that are constrained to integer values is called a mixed-integer linear programming problem. For modeling business problems, it is the preferred modeling approach in peer-review papers on sustainable supply chains and supply chain optimization (Mujkić et al. 2018).

7.1 *Generic Form of a MILP Problem*

A mixed integer linear programming problem can be expressed in the following generic form:

$$\text{Maximize } Z = c_1 \cdot x_1 + c_2 \cdot x_2 + \ldots c_i \cdot y_i + \ldots c_k \cdot y_k \ldots + c_n \cdot x_n \tag{2.26}$$

s.t.:

$$a_{11} \cdot x_1 + a_{12} \cdot x_2 + \ldots + a_{1i} \cdot y_i + \ldots a_{1k} \cdot y_k + \ldots + a_{1n} \cdot x_n \leq b_1 \tag{2.27}$$

$$a_{21} \cdot x_1 + a_{22} \cdot x_2 + \ldots + a_{2i} \cdot y_i + \ldots a_{2k} \cdot y_k + \ldots + a_{2n} \cdot x_n \leq b_2 \tag{2.28}$$

$$\ldots$$

$$a_{m1} \cdot x_1 + a_{m2} \cdot x_2 + \ldots + a_{mi} \cdot y_i + \ldots a_{mk} \cdot y_k + \ldots + a_{mn} \cdot x_n \leq b_m \tag{2.29}$$

and

$$x_1, x_2, \ldots x_n, \ldots y_i, \ldots y_z \geq 0 \tag{2.30}$$

$$x_1, \ldots x_n \in \mathbb{R} \tag{2.31}$$

$$y_i, \ldots, y_k \in \mathbb{Z} \tag{2.32}$$

Similar to the standard form of a LP problem discussed in Sect. 2, the MILP problem has one objective function (2.26) that is either maximized or minimized subject to some constraints expressed in (2.27) – (2.32). A MILP problem is different because it contains additional decision variables $y_i, \ldots y_k$, that are integers as expressed in constraint (2.32).

Special cases are if all decision variables need to be integer or binary. If they all need to be integer, they are called a pure integer linear programming problem, abbreviated as ILP or IP. If they all need to be binary, they are called a binary or

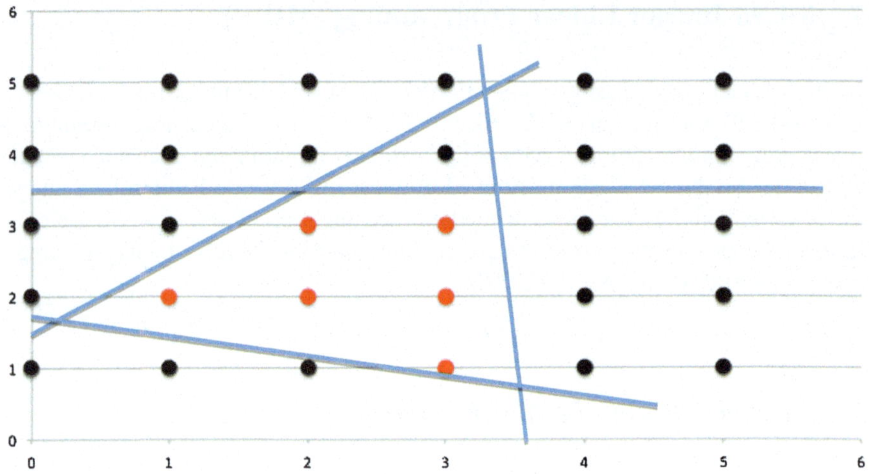

Fig. 2.18 The feasible region of an Integer Programming problem. Solutions in the feasible region are marked in red

Boolean linear programming problem, abbreviated as BIP. Both ILP and BIP are special cases of a MILP problem.

Solving a MILP problem is not just rounding the results to the next integer for the integer decision variables since it can return non-optimal and even non-feasible solutions. It requires a more complex algorithm. If the decision variables are only integer values, the feasible region differs from the feasible region discussed in Sect. 2.2. The feasible region of a pure ILP problem consists of only discrete integer values (see Fig. 2.18 indicated with red dots that lay within the feasible region indicated with the lines).

Solving a general MILP problem is NP-hard, meaning heuristic methods must be used. However, in some simple cases, exact solution algorithms such as the cutting plane method (i.e. first solving the problem as an LP problem then adding constraints so that only integer solutions are feasible) could be used. A commonly used algorithm is the branch and bound algorithm that we will not discuss within the scope of this book.

7.2 The Adapted MSW Allocation Problem Formulated as a MILP

To demonstrate how a MILP problem can be solved using the MS Excel Solver, we use the MSW problem discussed in the introduction of this chapter. Suppose that, because of a temporary problem, Incinerator A can only work at a maximum available capacity of 27.5 [ton/day]. Solving the LP problem of maximizing operational profit results in an optimal solution of $x_A = 27.5$ [ton/day] and $x_B = 31.67$ [ton/day] with an optimal operational result of 805 [€/day].

For planning ease, however, only positive integer values are allowed for the volumes to be incinerated in Incinerator A. Therefore, x_A is changed into the integer decision variables y_A.

This results in the adapted MILP problem below:

$$\text{Maximize } Z = 12 \cdot y_A + 15 \cdot x_B \ [€/\text{day}] \tag{2.33}$$

s.t.

$$0.2 \cdot y_A + 0.3 \cdot x_B \leq 15 \ [\text{ton/day}] \tag{2.34}$$

$$y_A + x_B \leq 80 \ [\text{ton/day}] \tag{2.35}$$

$$y_A \leq 30 \ [\text{ton/day}] \tag{2.36}$$

$$x_B \leq 40 \ [\text{ton/day}] \tag{2.37}$$

$$y_A, x_B \geq 0 \ [\text{ton/day}] \tag{2.38}$$

$$y_A \in \mathbb{Z} \tag{2.39}$$

$$x_B \in \mathbb{R} \tag{2.40}$$

A MILP problem can be easily solved using the MS Excel Solver. In Excel, the integer decision variable y_A (cell C10) has to be constrained as an integer in the MS Excel Solver dialogue box (see Fig. 2.19).

The outcome of this MILP problem is $y_A = 27$ [ton/day] and $y_B = 32$ [ton/day] and the total operational profit becomes 804 [€/day] (Fig. 2.20).

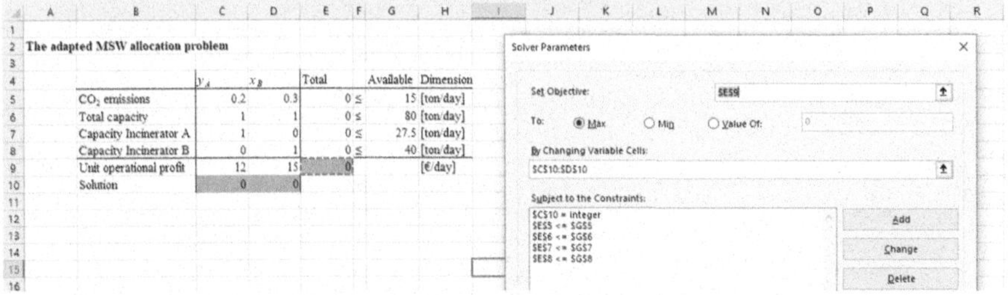

Fig. 2.19 The adapted MSW problem formulated as a MILP with integer decision variable y_A and real decision variable x_B

	A	B	C	D	E	F	G	H	I
1									
2	**The adapted MSW allocation problem**								
3									
4			y_A	x_B	Total		Available	Dimension	
5		CO_2 emissions	0.2	0.3	15	≤	15	[ton/day]	
6		Total capacity	1	1	59	≤	80	[ton/day]	
7		Capacity Incinerator A	1	0	27	≤	27.5	[ton/day]	
8		Capacity Incinerator B	0	1	32	≤	40	[ton/day]	
9		Unit operational profit	12	15	804			[€/day]	
10		Solution	27	32					
11									

Fig. 2.20 Outcome of the adapted MSW allocation problem

Chapter 3
Multiple Objective Optimization

Many sustainability problems consist of finding tradeoffs between various objectives subjected to a set of constraints. A well-known example is the People-Planet-Profit optimization. As soon as more than one objective function has to be optimized simultaneously, subjected to a same set of constraints, a problem is called a multi-objective optimization problem (MOOP).

In this chapter, we discuss the differences between a MOOP and a SOOP and present the weighted sum approach and the ε-constraint method to solve convex MOOPs formulated as multi-objective linear optimization problems (MOLPs). Moreover, we highlight the fundamental principles of multi-objective optimization and present a way to explore the Pareto optimal front, a function that contains all the optimal solutions for the MOOP.

We introduce multi-objective programming using a simple example of a bi-objective linear programming problem. Therefore, we simultaneously optimize two objectives of the Green Waste Valorization case.

1 Introduction: Bi-objective Optimization – The Green Waste Valorization Case

Let's consider the problem of the Green Waste Valorization case from Chap. 1 once again. Suppose we aim for the simultaneous maximization of the valorization of green waste as compost, expressed in objective function Z_1, and as a source of renewable energy, expressed in objective function Z_2. We consider the case of 1 ton of incoming green waste composed of 0.6 ton of wooden cuttings (Eq. 3.3), 0.2 ton of grass (Eq. 3.4) and 0.2 ton of leaves (Eq. 3.5) that can be composted and/or incinerated. The question to be solved for this MOOP is to find the optimal solution (s) that satisfy both objectives subjected to some constraints expressed in (3.3) – (3.7) and discussed in Sect. 6 of Chap. 1. This problem can be formulated as follows:

D. Inghels, *Introduction to Modeling Sustainable Development in Business Processes*, https://doi.org/10.1007/978-3-030-58422-1_3

$$\text{Maximize } Z_1 = (x_{wc} + x_{gc} + x_{lc}) \cdot 0.35 \tag{3.1}$$

$$\text{Maximize } Z_2 = 11.11 \cdot x_{wi} + 5.35 \cdot x_{gi} + 7.20 \cdot x_{li} \tag{3.2}$$

Subject to:

$$x_{wc} + x_{wi} = 0.6\,[\text{ton}] \tag{3.3}$$

$$x_{gc} + x_{gi} = 0.2\,[\text{ton}] \tag{3.4}$$

$$x_{lc} + x_{li} = 0.2\,[\text{ton}] \tag{3.5}$$

$$38.41 \cdot x_{wc} + 47.20 \cdot x_{gc} - 162.45 \cdot x_{lc} = 0 \tag{3.6}$$

$$x_{wc}, x_{gc}, x_{lc}, x_{wi}, x_{gi}, x_{li} \geq 0 \tag{3.7}$$

We first explore this two-objective optimization problem by solving the two single objective optimization problems Z_1 and Z_2 separately. Only maximizing Composting, i.e. Maximize Z_1 as expressed in (3.1) subjected to the constraints (3.3) – (3.7), results in the optimal objective values $z_1 = 0.35$ [ton] and $z_2 = 0$ [GJ]. Contrary, only maximizing Energy valorization Z_2, i.e. Maximize Z_2 as expressed in (3.2), results in the optimal objective values $z_1 = 0$ [ton] and $z_2 = 9.18$ [GJ].

Both results are two extreme optimal solutions of the MOOP formulated in (3.1) – (3.7). Many other solutions exist and lie between these two extreme optimal solutions; they are tradeoffs between the allocation of green waste to Compost and Energy. Several such equal optimal solutions are depicted in Fig. 3.1. The set of these tradeoff solutions is called the *Pareto optimal front,* and all these tradeoff solutions are called *Pareto optimal solutions*. All Pareto optimal solutions are equally optimal, but they differ in the sense that one solution is better in terms of one objective set.

The general outcome of a multi-objective optimization problem is a *Pareto optimal front,* also known as the *tradeoff curve* or *efficient frontier.* The Pareto optimal front is characterized by:

- A solution x is called Pareto optimal if no other solution is at least as good as x with respect to every objective and is strictly better than x with respect to at least one objective.
- A feasible solution y dominates a feasible solution x in a multi-objective problem if y is at least as good as x on every objective and is strictly better than x on at least one objective.
- Pareto optimal solutions are feasible solutions that are not dominated.

A Pareto optimal front represented by the Pareto curve can be convex, concave, or a combination of both. A Pareto curve is said to be convex if every point of a line segment connecting two arbitrary points on the Pareto curve lies "above" or on the Pareto curve (see Fig. 3.2). The Pareto curve is indicated in bold.

According to Deb (2009; definition 2.2), a multi-objective optimization problem is convex if all objective functions are convex, and the feasible region is convex.

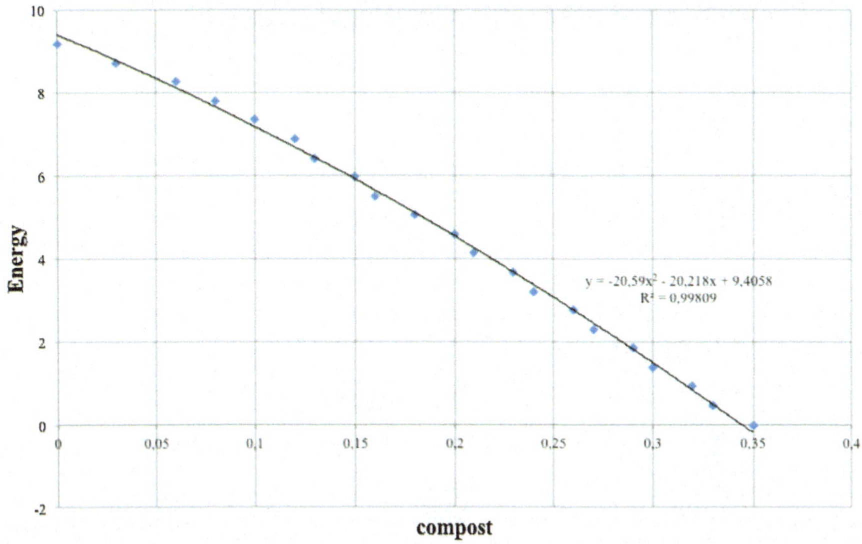

Fig. 3.1 Pareto optimal front of the bi-objective Green Waste Valorization case

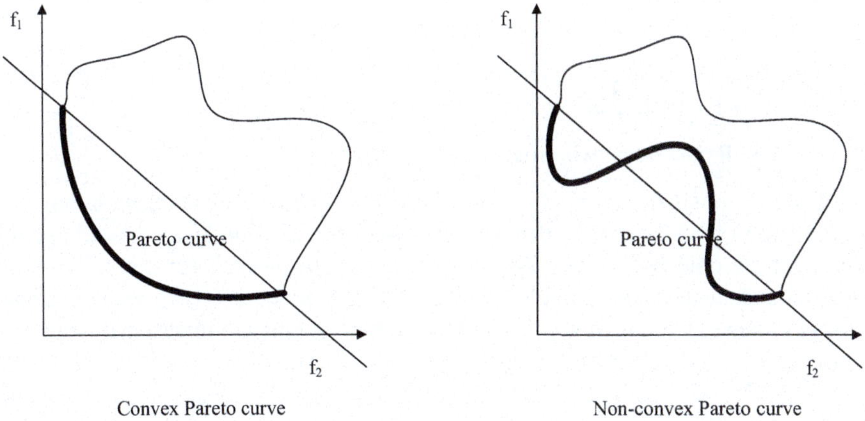

Fig. 3.2 Convex and non-convex Pareto curve

Linear functions are both convex and concave at the same time. Difficulties may be faced solving nonconvex MOOPs. Practically speaking, in the case of concave Pareto fronts, solving the MOOP will tend to give only extremal solutions, that is, solutions that are optimal in one of the objectives (Emmerich and Deutz 2018).

Two questions immediately arise: (i) how to determine the Pareto optimal front, and (ii) is it possible to derive one single optimal solution from the Pareto optimal front for a given MOOP? To answer these questions, we will first discuss the difference between a SOOP and MOOP.

2 The Difference Between a SOOP and a MOOP

Compared with a SOOP, a MOOP has the following additional properties (Burke and Kendall 2014):

(a) The cardinality of the optimal set is usually more than one.

The bi-objective Green Waste Valorization example in Sect. 1 clearly shows that this optimization problem with two conflicting objectives results in several Pareto optimal solutions. This applies to all MOOPs, no matter how many objectives are involved. By contrast, a SOOP generally results in one single optimal solution associated with this SOOP.

(b) There are two distinct goals of optimization instead of one.

A SOOP has only one goal: determining the optimal solution. A MOOP, however, has two distinct goals:

- Convergence to the Pareto optimal solutions, i.e. solving a MOOP should generate a set of Pareto optimal solutions
- Maintenance of a set of maximally spread Pareto optimal solutions, i.e. the set of Pareto optimal solutions should cover the entire region the Pareto optimal front associated with the MOOP and all these Pareto optimal solutions should be more or less equally spread over the entire Pareto optimal front

Each type of optimization algorithm for a MOOP should contain specific properties to achieve both goals.

(c) A MOOP possesses two different search spaces.

Contrary to single objective optimization, multi-objective optimization has two spaces (see Fig. 3.3): one for the decision variables, *Decision Space D*, and one for the multiple objective values, *Objective Space Z*. In general, for a problem with n independent decision variables and m objective functions, each n-dimensional vector x in the Decision Space is related to a point z in the m-dimensional Objective Space: $f(x_1, \ldots, x_n) = (z_1, z_2, \ldots, z_m)^T$.

3 The Standard Form of a MOOP

In general, a multi-objective optimization problem (MOOP) has several objective functions, which are to be minimized or maximized (please note that, in general, combinations of minimization and maximization are possible) and several constraints that any feasible solution, including the optimal solution, must satisfy.

A MOOP with M objective functions $f_m(x)$, J inequality constraints $g_j(x)$, K equality constraints $h_k(x)$ and n decision variables x_i (i = 1, 2, …, n) with lower bounds $x_{i,L}$ and upper bounds $x_{i,U}$ is formulated as follows (Deb 2009):

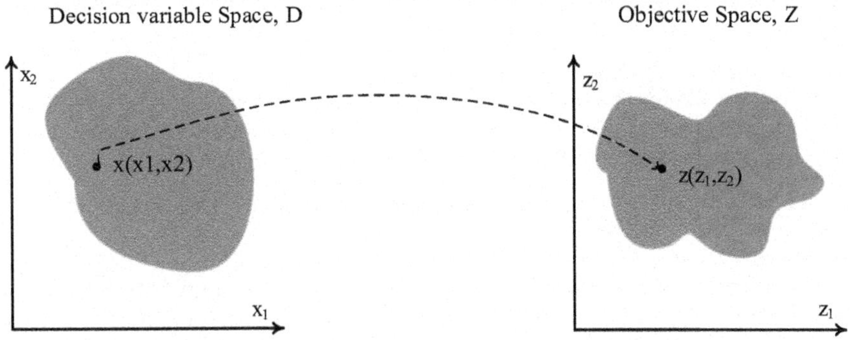

Fig. 3.3 The two spaces of a MOOP

$$
\begin{array}{llll}
\text{Minimize/Maximize} & f_m(x), & \mathrm{m} = 1,2,\ \ldots,\mathrm{M}; \\
\text{Subject to} & g_j(x) \geq 0, & \mathrm{j} = 1,2,\ \ldots,\mathrm{J}; \\
 & h_k(\mathrm{x}) = 0, & \mathrm{k} = 1,2,\ \ldots,\mathrm{K}; \\
 & x_{i,\mathrm{L}} \leq x_i \leq x_{i,U} & \mathrm{i} = 1,2,\ \ldots,\mathrm{n}.
\end{array}
$$

A solution x is a vector of n decision variables $x = (x_1, x_2, \ldots, x_n)^{\mathrm{T}}$. The bounds of decision variable x constitute the Decision Space, D. Note that similar to the objective functions and constraints of a SOOP, the minimization of objective functions can be transformed into a maximization objective function by multiplying by -1 and vice versa. The same applies to "greater than or equal to" or "less than or equal to" constraints.

A solution that does not satisfy all the J + K constraints and all the variable bounds is called an *infeasible solution*. In contrast, any solution x that satisfies all constraints and variable bounds is called a *feasible solution*. The set of all feasible solutions is called the *feasible region*, also known as the Search Space.

In this section, we only discussed linear objective functions and constraints. Whenever an objective or constraint function is nonlinear, the resulting problem is called a nonlinear multi-objective problem. We will not be discussing these types of sustainability problems within the scope of this book.

4 Two Approaches to Obtain One Single Solution in Multi-objective Optimization

In practice, users of MOOPs usually desire one single outcome, although a MOOP has a set of equal optimal outcomes that are gathered on the Pareto optimal front. Obtaining one single solution is achieved by adding decision-making information to the problem. This decision-making information is often expert or policy-driven and reflects a preference by the decision maker(s).

There are two approaches to determine one optimal solution of the Pareto optimal front (Narzisi 2008): the *preference-based procedure* and the *ideal procedure*.

The preference-based procedure (see Fig. 3.4) composites the objective functions as the weighted sum of these objectives. A relative preference vector $(w_1, w_2, \ldots, w_n)$ with $\sum_{i=1}^{n} w_i = 1; \forall w_i \geq 0$ is linked to the multiple objectives. This procedure is only useful if a relative preference vector for the objectives in known in advance.

The preference-based method has some advantages. It is simple and easy to use, and for convex problems, it finds solutions on the entire Pareto optimal set. However, this method also has some disadvantages. For mixed optimization problems with both minimization and maximization of objectives, one needs to convert all the objectives into one type, i.e. all objectives should be minimized or maximized. Next, a uniformly distributed set of weights does not guarantee a uniformly distributed set of Pareto optimal solutions, and two different sets of weight vectors will not necessarily lead to two different optimal solutions.

The ideal multi-objective optimization procedure (see Fig. 3.5) consists of a two-step approach (Burke and Kendall 2014):

Step 1: Find multiple tradeoff optimal solutions with a wide range of values for objectives.

Step 2: Choose one of the obtained solutions using a higher-level decision criterion.

This method may be more cumbersome than the preference-based method, but it does not have the disadvantages associated with this method.

We explain both methods in more detail in the next sections.

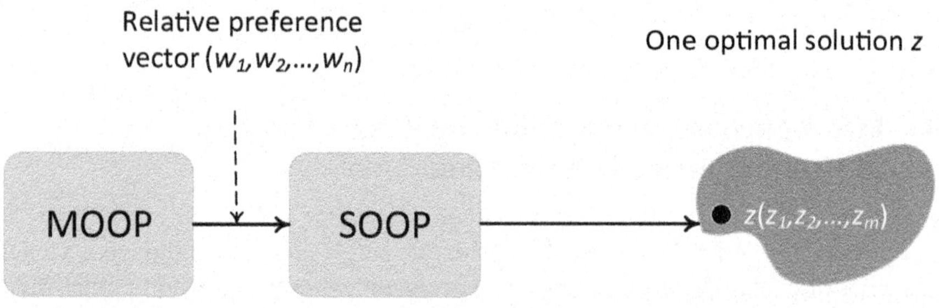

Fig. 3.4 Finding one optimal solution for a MOOP using the preference-based procedure

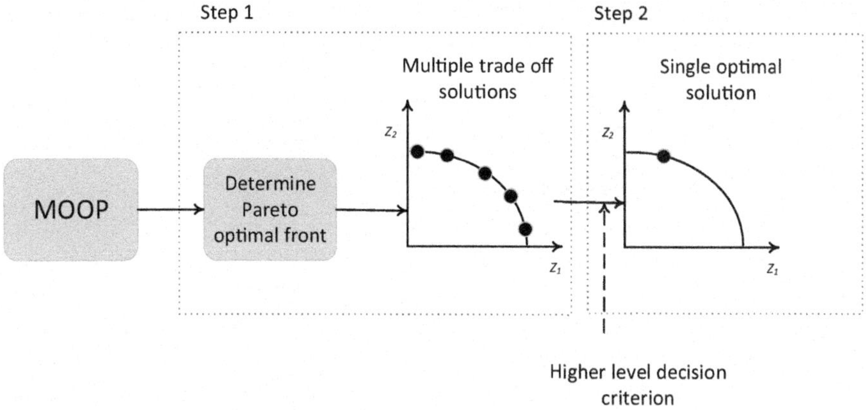

Fig. 3.5 Finding one optimal solution for a MOOP using the ideal procedure

5 The Preference-Based Procedure

We demonstrate the preference-based procedure using the example of the Green Waste Valorization case expressed as a bi-objective optimization problem in Sect. 1 of this chapter. Using the preference-based procedure, the objective functions Z_1 and Z_2 are grouped into one single objective function $Z_{1,2}$. More specifically, both objectives (3.1) and (3.2) are combined into one objective (3.8) using a relative preference vector (w_1, w_2):

$$\text{Maximize } Z_{1,2} = w_1 \cdot \left(x_{wc} + x_{gc} + x_{lc}\right) \cdot 0.35 + w_2 \cdot \left(11.11 \cdot x_{wi} + 5.35 \cdot x_{gi} + 7.20 \cdot x_{li}\right) \quad (3.8)$$

s.t. constraints (3.3)–(3.7)

$$\text{and with } w_1 + w_2 = 1; w_1, w_2 \geq 0 \quad (3.9)$$

The resulting SOOP can be solved using the MS Excel Solver, as explained in Chap. 2. First, we have to organize the objective functions by combining objective function Z_1, for Compost, and Z_2, for Energy, in the preference-based objective denoted "Compost and Energy obj." in cell B11 of Fig. 3.6. The formula for this combined objective function (3.8) is expressed in cell I11 = D14*I9 + D15*I10 (Fig. 3.6) with cells D14 and D15 containing the values w_1 and w_2 and cells I9 and I10 containing the expressions for Z_1 and Z_2 respectively.

The bi-objective Green Waste Valorization problem is now ready to be solved as a SOOP for any values of the weighting factors inserted in cells D14 and D15. For convenience, we have derived w_2 from the w_1 inserted value taking (3.9) into account (see cells D14 and D15). To find the optimal solution, we fill in the Solver dialogue box as depicted in Fig. 3.7. Note that only the objective (3.8) is used in the input field "Set Objective." It connects the individual objectives Z_1 and Z_2 together

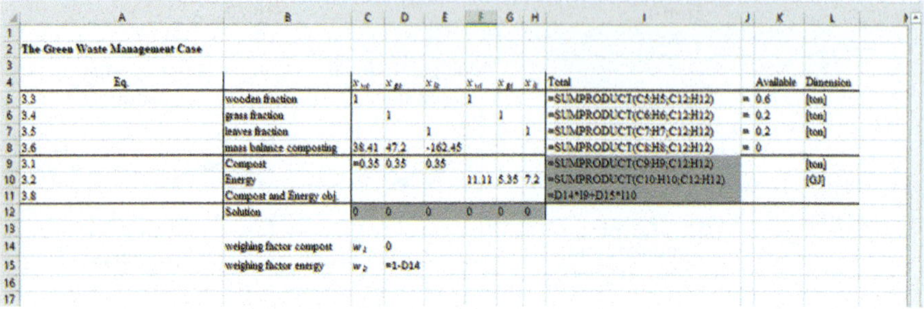

Fig. 3.6 Weighted sum approach in MS Excel for the Green Waste Management case

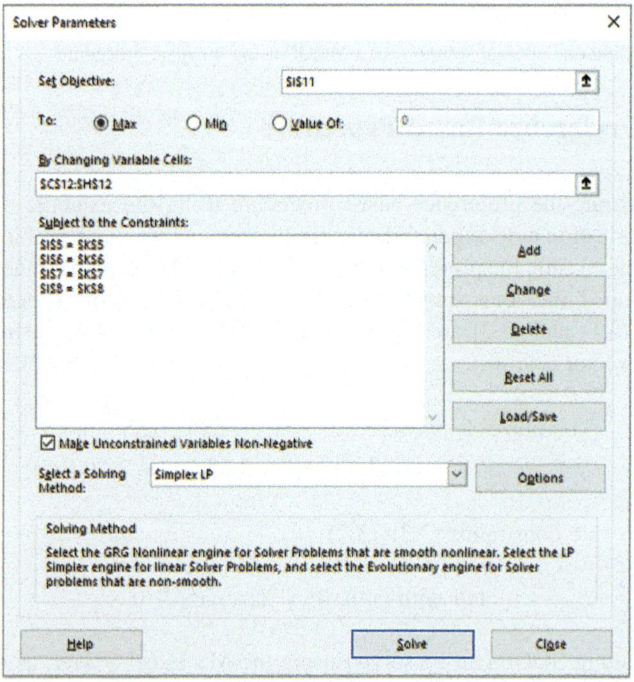

Fig. 3.7 The associated Solver Parameters for the weighted sum bi-objective optimization problem of the Green Waste Management problem

with the preference vector (w_1, w_2) in cell I11. The decision variables are defined in the range [C12:H12].

The optimal solution of this SOOP will depend on the values of the preference vector (w_1, w_2). We illustrate this for three cases: $w_1 = 1$ and $w_2 = 0$ (complete allocation to Compost valorization), $w_1 = 0$ and $w_2 = 1$ (complete allocation to Energy valorization), and $w_1 = w_2 = 0.5$ (equal preference for Compost and Energy allocation). Varying w_1 and w_2, and taking (3.9) into account, results in a set of optimal solutions that belong to the Pareto optimal front (Fig. 3.8).

	A	B	C	D	E	F	G	H	I	J	K	L	M
1													
2	**The Green Waste Valorization Case**												
3													
4	Eq.		x_{wc}	x_{gc}	x_{lc}	x_{wi}	x_{gi}	x_{li}	Total		Available	Dimension	
5	3.3	wooden cuttings	1			1			0.60	=	0.6	[ton]	
6	3.4	grass		1			1		0.20	=	0.2	[ton]	
7	3.5	leaves			1			1	0.20	=	0.2	[ton]	
8	3.6	mass balance composting	38.41	47.2	-162.45				0.00	=	0		
9	3.1	Compost objective	0.35	0.35	0.35				0.00			[ton]	
10	3.2	Energy objective				11.11	5.35	7.2	9.18			[GJ]	
11	3.8	Compost and Energy obj.							9.18				
12		Solution	0.00	0.00	0.00	0.60	0.20	0.20					
13													
14		weighing factor compost	w_1	0									
15		weighing factor energy	w_2	1									

	A	B	C	D	E	F	G	H	I	J	K	L	M
1													
2	**The Green Waste Valorization Case**												
3													
4	Eq.		x_{wc}	x_{gc}	x_{lc}	x_{wi}	x_{gi}	x_{li}	Total		Available	Dimension	
5	3.3	wooden cuttings	1			1			0.60	=	0.6	[ton]	
6	3.4	grass		1			1		0.20	=	0.2	[ton]	
7	3.5	leaves			1			1	0.20	=	0.2	[ton]	
8	3.6	mass balance composting	38.41	47.2	-162.45				0.00	=	0		
9	3.1	Compost objective	0.35	0.35	0.35				0.35			[ton]	
10	3.2	Energy objective				11.11	5.35	7.2	0.00			[GJ]	
11	3.8	Compost and Energy obj.							0.35				
12		Solution	0.60	0.20	0.20	0.00	0.00	0.00					
13													
14		weighing factor compost	w_1	1									
15		weighing factor energy	w_2	0									
16													

	A	B	C	D	E	F	G	H	I	J	K	L	M
1													
2	**The Green Waste Valorization Case**												
3													
4	Eq.		x_{wc}	x_{gc}	x_{lc}	x_{wi}	x_{gi}	x_{li}	Total		Available	Dimension	
5	3.3	wooden cuttings	1			1			0.60	=	0.6	[ton]	
6	3.4	grass		1			1		0.20	=	0.2	[ton]	
7	3.5	leaves			1			1	0.20	=	0.2	[ton]	
8	3.6	mass balance composting	38.41	47.2	-162.45				0.00	=	0		
9	3.1	Compost objective	0.35	0.35	0.35				0.00			[ton]	
10	3.2	Energy objective				11.11	5.35	7.2	9.18			[GJ]	
11	3.8	Compost and Energy obj.							4.59				
12		Solution	0.00	0.00	0.00	0.60	0.20	0.20					
13													
14		weighing factor compost	w_1	0.5									
15		weighing factor energy	w_2	0.5									

Fig. 3.8 Solutions of the bi-objective GWM problem: (**a**) if $w_1 = 0$ and $w_2 = 1$; (**b**) if $w_1 = 1$ and $w_2 = 0$ and (**c**) if $w_1 = w_2 = 0.5$

The optimal solution for $w_1 = 0$ and $w_2 = 1$ (full preference for Energy allocation) results in an optimal solution of the decision variables (x_{wc}, x_{gc}, x_{lc}, x_{wi}, x_{gi}, x_{li}) = (0, 0, 0, 0.60, 0.20, 0.20) and the complete allocation of green waste to Energy. In contrast, the optimal solution for $w_1 = 1$ and $w_2 = 0$ (full preference for Compost allocation) results in an optimal solution of the decision variables (x_{wc}, x_{gc}, x_{lc}, x_{wi}, x_{gi}, x_{li}) = (0.60, 0.20, 0.20, 0, 0, 0) and the complete allocation of green waste to Compost.

The optimal solution for $w_1 = w_2 = 0.5$ (equal preference for Energy and Compost allocation) results in an optimal solution of the decision variables (x_{wc}, x_{gc}, x_{lc}, x_{wi}, x_{gi}, x_{li}) = (0, 0, 0, 0.60, 0.20, 0.20), which is identical to the complete allocation of green waste to Energy. This may not correspond to what is intuitively expected. However, it is an illustration that the size of the weighting factors does not necessarily correspond with the position on the Pareto optimal front, as discussed in Sect. 4.

6 The Ideal Procedure

As introduced in Sect. 4 of this chapter, the ideal procedure consists of a two-step approach. First, the Pareto optimal front has to be determined then one value on this Pareto optimal front has to be selected using higher-level information. We demonstrate this procedure using the Green Waste Valorization bi-objective optimization problem outlined in Sect. 1 of this chapter.

6.1 *Step 1- Determining the Pareto-optimal Front*

The first step of the ideal procedure is to determine the Pareto optimal front. For the bi-objective function, the Pareto optimal front is limited between the full allocation of green waste to Compost or Energy. This corresponds to (z_1, z_2) = (0.35, 0) and (z_1, z_2) = (0, 9.18) respectively. These two points are the extreme points on the Pareto optimal front in the Objective Space, Z. All other points on the Pareto optimal front are situated between these two extreme points. The Pareto optimal front can be derived by selecting any values of z_1 or z_2 within the allowable range between the two extreme points on the Pareto optimal front and by calculating the corresponding values for z_2 and z_1.

To illustrate this, let's consider a random data point for the Energy objective $z_2 = 4.130$ GJ that belongs to the Pareto optimal front (see Fig. 3.1 and Table 3.1) Now let's calculate the corresponding Compost objective value z_1 by assigning the right-hand side value $z_2 = 4.130$ [GJ] for Energy to the function Z_2. Next, we make this objective a new constraint (3.10) for the MOOP by assigning it a right-hand side value so that the bi-objective optimization problem is transformed into a single objective optimization problem.

This transforms the MOOP to a SOOP as follows:

$$\text{Maximize } Z_1 = \left(x_{wc} + x_{gc} + x_{lc}\right) \cdot 0.35 \tag{3.1}$$

Subject to:

$$x_{wc} + x_{wi} = 0.6\ [\text{ton}] \tag{3.3}$$

$$x_{gc} + x_{gi} = 0.2\ [\text{ton}] \tag{3.4}$$

$$x_{lc} + x_{li} = 0.2\ [\text{ton}] \tag{3.5}$$

$$38.41 \cdot x_{wc} + 47.20 \cdot x_{gc} - 162.45 \cdot x_{lc} = 0 \tag{3.6}$$

$$11.11 \cdot x_{wi} + 5.35 \cdot x_{gi} + 7.20 \cdot x_{li} = 4.13 \tag{3.10}$$

$$x_{wc}, x_{gc}, x_{lc}, x_{wi}, x_{gi}, x_{li} \geq 0 \tag{3.7}$$

The outcome of this SOOP is depicted in Fig. 3.9. This figure shows that the corresponding optimal value for Compost is $z_1 = 0.21$ ton, which is the corresponding value for Compost on the Pareto optimal front as depicted in Fig. 3.1 and Table 3.1.

We can repeat this for a set of numbers z_2 belonging to the range [0, 9.18] [GJ]. This would result in a cumbersome number of SOOPs to be optimized. Instead of calculating all these points separately, the *MS Excel Solver Table* saves a lot of work. This MS Excel add-in executes all the separate calculations in one go.

We briefly explain how the MS Excel Solver works. First, the MS Excel Solver Table needs to be installed as an add-in and activated. Please consult MS Excel for more information on how to do this for the version and operating system of your PC. Once installed and activated, it will show up as a separate tab in Excel (see Fig. 3.10).

Next, the MOOP has to be re-written as a single objective problem by making one of the objectives a constraint. We opt to do this for the Z_2 objective following the formulation (3.1) – (3.10) described above. Moreover, a separate table has to be added that contains the data points for the Energy objective values belonging to the Pareto optimal front (see Fig. 3.11). The goal of this table is to assign the RHS value of (3.10) expressed in cell K8 different values within the allowable range of [0, 9.18] [GJ]; the corresponding z_1 value and decision variables will be calculated when the Solver Table is executed. In Fig. 3.11, we have opted to calculate the Pareto optimal front for 9 points equally distributed over $z_2 = 1$ to $z_2 = 9$ per step of 1 [GJ]. The Excel sheet row 14 contains the values that will be calculated for each of the nine z_2 values listed up in the cells [B15:B23].

Table 3.1 Data points in the Pareto optimal front

Data point	**1**	**2**	**3**	**4**	**5**	**6**	**7**	**8**	**9**	**10**	**11**	**12**
Compost z_1 [ton]	0.000	0.030	0.060	0.080	0.100	0.120	0.130	0.150	0.160	0.180	0.200	0.210
Energy z_2 [GJ]	9.180	8.720	8.260	7.800	7.340	6.880	6.420	5.960	5.510	5.050	4.590	4.130
Data point	**13**	**14**	**15**	**16**	**17**	**18**	**19**	**20**	**21**			
Compost z_1 [ton]	0.230	0.240	0.260	0.270	0.290	0.300	0.320	0.330	0.350			
Energy z_2 [GJ]	3.670	3.210	2.750	2.290	1.840	1.380	0.920	0.460	0.000			

The Green Waste Management Case											
Eq.		x_{wc}	x_{gc}	x_{lc}	x_{wi}	x_{gi}	x_{li}	Total		Available	Dimension
3.3	wooden fraction	1			1			0.60	=	0.6	[ton]
3.4	grass fraction		1			1		0.20	=	0.2	[ton]
3.5	leaves fraction			1			1	0.20	=	0.2	[ton]
3.6	mass balance composting	38.41	47.2	-162.45				0.00	=	0	
3.2	Energy				11.11	5.35	7.20	4.13	=	4.13	[GJ]
3.1	Compost	0.35	0.35	0.35				0.21			[ton]
	Solution	0.28	0.20	0.12	0.32	0.00	0.08				

Fig. 3.9 Calculation of z_1 for $z_2 = 4.13$

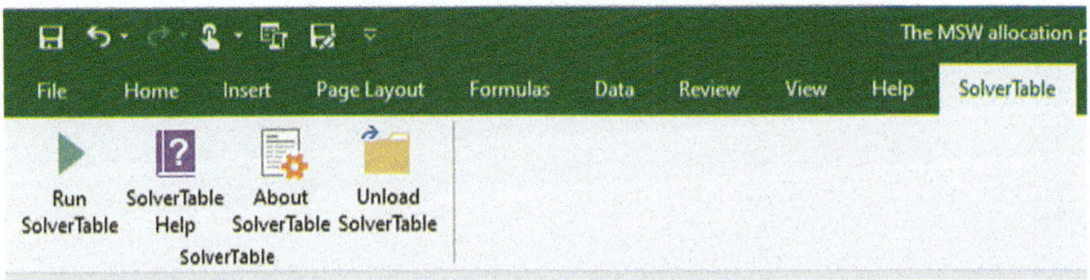

Fig. 3.10 Solver Table add-in, once installed and activated in MS Excel

	A	B	C	D	E	F	G	H	I	J	K	L
1	The Green Waste Valorization Case											
2												
3	Eq.		x_{wc}	x_{gc}	x_{lc}	x_{wi}	x_{gi}	x_{li}	Total		Available	Dimension
4	3.3	wooden cuttings	1			1			=SUMPRODUCT(C4:H4;C10:H10)	=	=0.6	[ton]
5	3.4	grass		1			1		=SUMPRODUCT(C5:H5;C10:H10)	=	=0.2	[ton]
6	3.5	leaves			1			1	=SUMPRODUCT(C6:H6;C10:H10)	=	=0.2	[ton]
7	3.6	mass balance composting	38.41	47.2	-162.45				=SUMPRODUCT(C7:H7;C10:H10)	=	0	
8	3.2	Energy				11.11	5.35	7.2	=SUMPRODUCT(C8:H8;C10:H10)	=	9.18	[GJ]
9	3.1	Compost	=0.35	0.35	0.35				=SUMPRODUCT(C9:H9;C10:H10)			[ton]
10		Solution	0	0	0	0	0	0				
11												
12												
13		Energy yield	x_{wc}	x_{gc}	x_{lc}	x_{wi}	x_{gi}	x_{li}	*Compost*			
14			=C10	=D10	=E10	=F10	=G10	=H10	=I9			
15		1										
16		2										
17		3										
18		4										
19		5										
20		6										
21		7										
22		8										
23		9										

Fig. 3.11 Preparing the Green Waste Valorization case for using the Solver Table

Before executing the Oneway Solver Table function, one should first solve the SOOP with the MS Excel Solver, as explained in Sect. 2. The result is depicted in Fig. 3.12.

	A	B	C	D	E	F	G	H	I	J	K	L	M
1	The Green Waste Valorization Case												
2													
3	Eq.		x_{wc}	x_{gc}	x_{lc}	x_{wi}	x_{gi}	x_{li}	Total		Available	Dimension	
4	3.3	wooden cuttings	1			1			0.60	=	0.60	[ton]	
5	3.4	grass		1			1		0.20	=	0.20	[ton]	
6	3.5	leaves			1			1	0.20	=	0.20	[ton]	
7	3.6	mass balance composting	38.41	47.2	-162.45				0.00	=	0.00		
8	3.2	Energy				11.11	5.35	7.20	9.18	=	9.18	[GJ]	
9	3.1	Compost	0.35	0.35	0.35				0.00			[ton]	
10		Solution	0.00	0.00	0.00	0.60	0.20	0.20					
11													
12													
13		Energy yield	x_{wc}	x_{gc}	x_{lc}	x_{wi}	x_{gi}	x_{li}	*Compost*				
14			0.00	0.00	0.00	0.60	0.20	0.20	0.00				
15		1											
16		2											
17		3											
18		4											
19		5											
20		6											
21		7											
22		8											
23		9											

Fig. 3.12 Solution of the SOOP denoted in Fig. 3.11

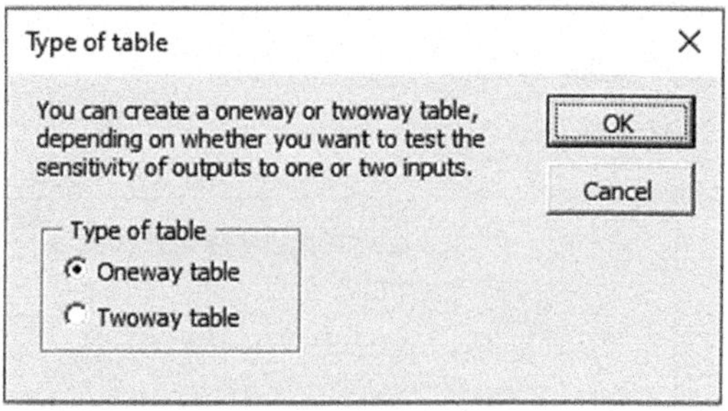

Fig. 3.13 Selecting the Oneway MS Excel Solver Figure

The problem is now organized correctly to execute the Oneway Solver Table. When the Solver Table add-in is launched, the "Type of table" selection menu opens. Select the Oneway table radio button then click the "OK" button (see Fig. 3.13).

When the Oneway table is opened, the parameters for the Oneway table have to be entered. This involves assigning the RHS value that has to be changed. This is the RHS value of constraint (3.10) expressed in cell K8 for the Green Waste Valorization case. Next, right-hand side values that will be assigned to cell K8 must be specified as a range. In this case, they are expressed in cells B15 – B23. Finally, the output cells for the problem have to be selected. In this case, these are the decision variables for the Green Waste Valorization problem and the objective function values z_1 for Compost. These values are assigned to the cells C14 – I14 (see Fig. 3.14).

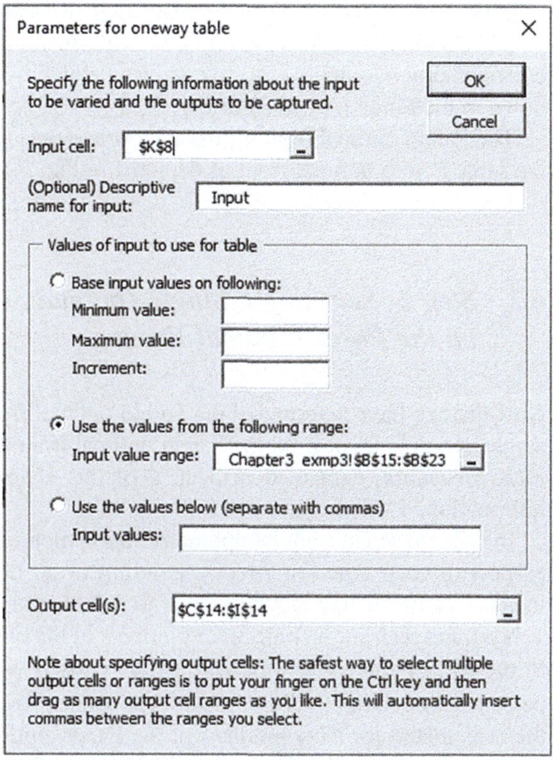

Fig. 3.14 Constructing the Oneway Solver Table for the Green Waste Valorization problem

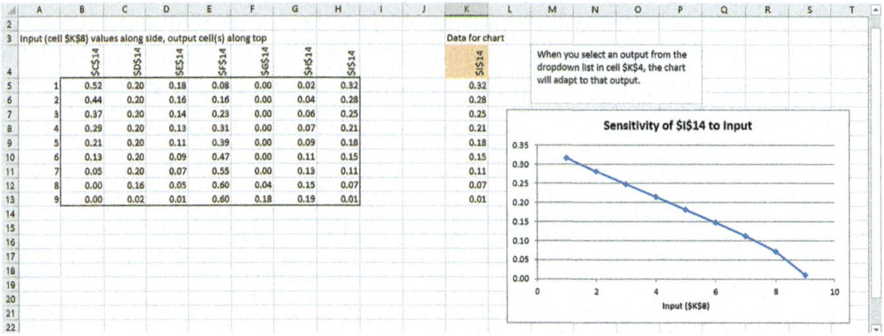

Input (cell $K$8) values along side, output cell(s) along top

	$C$14	$D$14	$E$14	$F$14	$G$14	$H$14	$I$14
1	0.52	0.20	0.18	0.08	0.00	0.02	0.32
2	0.44	0.20	0.16	0.16	0.00	0.04	0.28
3	0.37	0.20	0.14	0.23	0.00	0.06	0.25
4	0.29	0.20	0.13	0.31	0.00	0.07	0.21
5	0.21	0.20	0.11	0.39	0.00	0.09	0.18
6	0.13	0.20	0.09	0.47	0.00	0.11	0.15
7	0.05	0.20	0.07	0.55	0.00	0.13	0.11
8	0.00	0.16	0.05	0.60	0.04	0.15	0.07
9	0.00	0.02	0.01	0.60	0.18	0.19	0.01

Data for chart
$I$14
0.32
0.28
0.25
0.21
0.18
0.15
0.11
0.07
0.01

Fig. 3.15 Outcome of the Oneway Solver analysis of the Green Waste Valorization problem

Executing the Oneway Solver Table results in the calculation of the parameter values assigned to cells C14 – I14 (see Fig. 3.12). The Oneway Solver analysis outcome is placed on a separate tab in Excel (cells B4 – H4 in Fig. 3.15 refer to cells C4 – I14 in Fig. 3.12). The Compost objective function value z_I has been assigned to cell I14 (see Fig. 3.12). Therefore, cell I14 has to be selected from the dropdown list in cell K4 in Fig. 3.15 to get the corresponding Pareto optimal values for the Energy objective values located on the Pareto optimal front. Since we assigned cell I14 to Compost in Fig. 3.12, cell K4 has to be selected in Fig. 3.15 because it represents the corresponding Compost objective values belonging to the Pareto optimal front laid down in the range [A5:A13] of Fig. 3.15.

The Pareto optimal front is displayed below the table in Fig. 3.15. Note that this is the same Pareto optimal front as depicted in Fig. 3.1.

6.2 Step 2- Selecting a Single Optimal Solution on the Pareto-optimal Front

Now that we have determined the Pareto optimal front, it is possible to choose one single optimal outcome on the Pareto optimal front for this MOOP. We will use the ideal procedure, explained in Sect. 4 of this chapter, and must add higher-level information.

In order to select a single optimal solution on the Pareto optimal front, we assign ex-post to each criterion j (corresponding to an objective function), a weighting factor, v_j, representing its importance so that the total of all weighting factors equals 1. Next, for each single objective solution s of the optimal solutions set, the distance to the ideal point z_j^* is weighted by the chosen weights z_j^s to obtain a weighted percentage deviation (WPD_s) factor. The value z_j^s of the jth objective function Z_j is then calculated for all solutions s of the Pareto optimal set, and compared with the ideal point value of the jth objective function z_j^*. This ideal point value is formed by the optimal results per objective function of each assessed alternative depicted in the payoff table. The solution s with the lowest WPD_s value is the optimal solution since it is closest to the ideal point. The payoff table contains the extreme points on the Pareto optimal front. For a bi-objective MOOP ($m = 2$), the WPD_s factor is expressed in (3.12).

$$WPD_s = \sum_{j=1}^{m=2} v_j \cdot \left[\frac{|z_j^s - z_j^*|}{z_j^*} \right]; v_1 + v_2 = 1;$$
$$v_1, v_2 \geq 0 \tag{3.12}$$

Applied to the Pareto optimal front population of the Green Waste Valorization case, the payoff table is depicted in Table 3.2.

Using (3.12), the single optimal solution for the Green Waste Valorization problem depends on the assigned expert value for v_1 and v_2 as depicted in

Table 3.2 Payoff table Green Waste Valorization problem

	Compost Z_1 [ton]	Energy Z_2 [GJ]
Max Z_1	0.350 (z^*_1)	0
Max Z_2	0	9.180 (z^*_2)

Table 3.3 preferred single optimal solution on the Pareto-optimal front for the Green Waste Valorization problem

		WPD_s		
Compost: z_1 [ton]	Energy: z_2 [GJ]	$v_{1,2} = 0.5$	$v_1 = 0.3$; $v_2 = 0.7$	$v_1 = 0;7$; $v_2 = 0.3$
0.350	0.000	0.500	0.700	**0.300**
0.316	1.000	0.494	0.653	0.335
0.282	2.000	0.488	0.605	0.370
0.249	3.000	0.481	0.558	0.405
0.215	4.000	0.475	0.511	0.439
0.181	5.000	0.469	0.463	0.474
0.147	6.000	0.463	0.416	0.509
0.114	7.000	**0.457**	0.369	0.544
0.071	8.000	0.462	0.329	0.596
0.011	9.000	0.495	0.305	0.685
0.000	9.180	0.500	**0.300**	0.700

Table 3.3. The lowest associated WPD_s value indicates the preferred optimal solution. If $v_1 = v_2 = 0.5$ then the optimal outcome is $(z_1, z_2) = (0.114, 7.000)$ on the Pareto optimal front. If Energy valorization is preferred over Compost ($v_1 = 0.3$ and $v_2 = 0.7$), then Green Waste is completely assigned to Energy valorization as the preferred solution.

7 The Augmented ε-Constraint Method

Using MS Excel for the preference-based method works well for bi-objective MOOPs. Many sustainability problems in business can be formulated as bi-objective MOOPs. However, sometimes more than two objectives have to be optimized in a MOOP.

If a MOOP consists of many objectives, the augmented ε-constraint method (AUGMECON), a widely used preference-based approach that results in the exact calculation of some points on the Pareto optimal front, can be used. The example discussed in this section is based on Inghels et al. (2019) to illustrate the ε-constraint method. The augmented ε-constraint method is a new version of the conventional ε-constraint method. It provides remedies for well-known pitfalls in the conventional ε-constraint method, such as generating weak Pareto optimal and redundant solutions (Mavrotas 2009). The AUGMECON method is available in several different modeling languages, including GAMS (general algebraic modeling language, www.gams.com). For a detailed explanation of the AUGMECON method, we refer interested readers to Mavrotas (2007, 2009).

In the ε-constraint method, one of the objective functions is optimized using the other objective functions as constraints. Let's look at the Green Waste Valorization case once again. We extend the bi-objective MOOP discussed in Sect. 1 with a third objective, Z_3, which minimizes the environmental impact discussed in Sect. 6 of Chap. 1. The MOOP to be solved is similar to the MOOP discussed in Sect. 1, but expanded with a third objective (3.13).

$$\text{Maximize} \quad Z_1 = \left(x_{wc} + x_{gc} + x_{lc}\right) \cdot 0.35 \tag{3.1}$$

$$\text{Maximize} \quad Z_2 = 11.11 \cdot x_{wi} + 5.35 \cdot x_{gi} + 7.20 \cdot x_{li} \tag{3.2}$$

$$\text{Minimize} \quad Z_3 = -575 \cdot \left(x_{wc} + x_{gc} + x_{lc}\right) - 1710 \cdot \left(x_{wi} + x_{gi} + x_{li}\right) \tag{3.13}$$

Subject to:

$$x_{wc} + x_{wi} = 0.6\,[\text{ton}] \tag{3.3}$$

$$x_{gc} + x_{gi} = 0.2\,[\text{ton}] \tag{3.4}$$

$$x_{lc} + x_{li} = 0.2\,[\text{ton}] \tag{3.5}$$

$$38.41 \cdot x_{wc} + 47.20 \cdot x_{gc} - 162.45 \cdot x_{lc} = 0 \tag{3.6}$$

$$x_{wc}, x_{gc}, x_{lc}, x_{wi}, x_{gi}, x_{li} \geq 0 \tag{3.7}$$

The objective function (3.1) to maximize Compost is selected as the objective to be optimized, and the two other linear objective functions (3.2) and (3.13) are used as constraints together with the other existing constraints. As a result, the set of constraints remains linear.

Since all objective functions must be written as maximization objectives, the minimization objective (3.13) is first transferred into a maximization objective using the duality principle *Minimize* Z_j = *Maximize* $(-Z_j)$ before it is used as a constraint. Therefore (3.13) becomes (3.14):

$$\text{Maximize} \quad Z_3 = 575 \cdot \left(x_{wc} + x_{gc} + x_{lc}\right) + 1710 \cdot \left(x_{wi} + x_{gi} + x_{li}\right) \tag{3.14}$$

The ε-constraint method forces a constrained problem to produce only efficient solutions by adding a factor to the first objective function (Mavrotas and Florios 2013; Mavrotas 2009). Furthermore, the ranges of the objective function are divided into equal intervals proportional to the number of grid points chosen. We opt for 10 points; therefore, each of the ranges is divided into ten equal intervals. The right-hand side value of the objective functions Z_2 *and* Z_3 are assigned a value e_2 *and* e_3. These e values range proportionally between the lowest and highest value in the payoff table for Z_2 *and* Z_3. Furthermore, the ranges of the respective objective functions Z_j are r_2, r_3; a range r_j is the difference between the highest and lowest value in the payoff table for Z_j. Finally, S_2, S_3 are the slack or surplus variables of the respective constraints and $\varepsilon \in [10^{-6}, 10^{-3}]$.

This results in the following linear programming model:

$$\text{Maximize } (x_{wc} + x_{gc} + x_{lc}) \cdot 0.35 + \varepsilon \cdot \left(\frac{S_2}{r_2} + \frac{S_3}{r_3}\right) \quad (3.15)$$

Subject to:

$$11.11 \cdot x_{wi} + 5.35 \cdot x_{gi} + 7.20 \cdot x_{li} - S_2 = e_2 \quad (3.16)$$

$$575 \cdot (x_{wc} + x_{gc} + x_{lc}) + 1710 \cdot (x_{wi} + x_{gi} + x_{li}) - S_3 = e_3 \quad (3.17)$$

$$x_{wc} + x_{wi} = 0.6\ [\text{ton}] \quad (3.18)$$

$$x_{gc} + x_{gi} = 0.2\ [\text{ton}] \quad (3.19)$$

$$x_{lc} + x_{li} = 0.2\ [\text{ton}] \quad (3.20)$$

$$38.41 \cdot x_{wc} + 47.20 \cdot x_{gc} - 162.45 \cdot x_{lc} = 0\ [\text{ton}] \quad (3.21)$$

$$x_{wc}, x_{gc}, x_{lc}, x_{wi}, x_{gi}, x_{li} \geq 0\ [\text{ton}] \quad (3.22)$$

Efficient solutions to the problem are obtained by parametric variation of the variables on the RHS of the second and third constrained objective functions (e_2 and e_3) (Mavrotas 2009). Solving the MOOP results in a payoff table that expresses the results of individual optimization of objective functions, as depicted in Table 3.4. According to Deb (2009), the optimal objective values of each single objective optimization model Z_j form the components of the ideal point vector z^*_j for the multi-objective problem under investigation.

We used the ε-constraint module of GAMS (www.gams.com) to derive the payoff table and to solve the problem formulated above. This module is based on linear programming.

The results of the GAMS optimization procedure with 10 grid points for e_i lead to the Pareto optimal front shown per objective; the selected points are shown on the left part of Table 3.5.

The next step in the preference-based method is to select one single optimal solution on the Pareto optimal front by adding the expert preference. We use the weighted percentage deviation (WPD_s) factor once again. In the case of three objectives, the WPD_s factors are calculated as in (3.15). The outcome for a set of options v_1, v_2, and v_3 linked to a preference for Z_1, Z_2, and Z_3, respectively, and using

Table 3.4 Payoff table for the Green Waste Valorization problem extended with objective Z_3

	Compost Z_1[ton]	Energy Z_2 [GJ]	Normalized environmental impact Z_3 [pt]
Max Z_1	**0.35** (z^*_1)	0	−575.00
Max Z_2	0	**9.18** (z^*_2)	−1710.00
Min Z_3	0	9.18	**−1710.00** (z^*_3)
Range r_2		9.18	
Range r_3			1135.00

Table 3.5 *WPD_s* values of the Green Waste Valorization problem for five sets of weighting factors

				WPD_s				
Nr *s*	Compost [ton]	WTE [GJ]	LCI [pt]	$v_{1,2,3} = 1/3$	$v_{1,2} = 0.45$, $v_3 = 0.10$	$v_1 = 0.6$, $v_{2,3} = 0.2$	$v_2 = 0.6$, $v_{1,3} = 0{,}2$	$v_3 = 0{,}6$, $v_{1,2} = 0.2$
1	0.00	9.18	−1710.00	**0.333**	0.450	0.600	**0.200**	**0.200**
2	0.03	8.60	−1596.50	0.348	0.446	0.574	0.234	0.235
3	0.06	8.26	−1529.40	0.345	0.429	0.538	0.247	0.249
4	0.07	8.02	−1483.01	0.353	0.430	0.532	0.262	0.265
5	0.10	7.25	−1369.51	0.375	0.436	0.510	0.309	0.304
6	0.10	7.34	−1379.06	0.369	**0.431**	0.507	0.302	0.299
7	0.13	6.42	−1278.56	0.394	0.443	0.488	0.357	0.337
8	0.14	6.22	−1256.01	0.396	0.442	0.478	0.367	0.344
9	0.16	5.51	−1178.05	0.418	0.455	0.468	0.411	0.375
10	0.17	5.18	−1142.51	0.427	0.461	0.462	0.431	0.389
11	0.20	4.59	−1077.55	0.433	0.455	0.431	0.460	0.408
12	0.21	4.14	−1029.02	0.449	0.467	0.429	0.489	0.429
13	0.23	3.67	−977.05	0.457	0.467	0.411	0.514	0.446
14	0.24	3.11	−915.52	0.480	0.485	0.414	0.553	0.474
15	0.26	2.75	−876.54	0.482	0.480	0.392	0.569	0.484
16	0.28	2.07	−802.02	0.502	0.492	0.381	0.611	0.513
17	0.29	1.84	−776.04	0.506	0.492	0.372	0.623	0.522
18	0.31	1.04	−688.53	0.533	0.510	0.365	0.674	0.559
19	0.32	0.92	−675.53	0.530	0.504	0.352	0.678	0.560
20	0.35	0.00	−575.03	0.555	0.516	**0.333**	0.733	0.598

The lowest *WPD_s* value, indicated in bold, corresponds to the preferred Compost and WTE valorization fractions the ideal points z^*_j from the payoff Table 3.4 is again expressed in a WPD_s score using (3.15). The lowest outcome, depicted in bold in Table 3.5, for a given set of preferences v_1, v_2, and v_3 is the single optimal solution linked to this set of preferences.

$$WPD_s = \sum_{j=1}^{m=3} v_j \cdot \left[\frac{|z_j^s - z_j^*|}{z_j^*} \right]; v_1 + v_2 + v_3 = 1; v_1, v_2, v_3 \geq 0 \quad (3.15)$$

Depending on the set of preference values, the optimal solution is situated at the extreme points of the Pareto optimal front or somewhere in between.

8 Other Solution Methods

In addition to the weighted sum method, many other methods can alleviate the difficulties faced by the weighted sum approach in solving problems with non-convex objective spaces. Other methods for solving MOOPs include the Value Function Method, Goal Programming Methods, and evolutionary algorithms. For more details, we refer to Deb (2009).

Chapter 4
Quantifying the Economic Impact

In the two previous chapters, we discussed how to optimize sustainability problems in the event of one objective or multiple conflicting objectives. Many sustainability problems in business processes can be formulated as a People-Planet-Profit problem. In Chaps. 4, 5 and 6, we shortly discuss some of the common practices to quantify the three pillars of sustainability. The output of each of the chapters can be used as an input in a SOOP or MOOP.

The economic pillar of the triple bottom line is commonly quantified by minimizing costs or maximizing profits (Seuring and Müller 2008). For supply chain optimization, minimizing costs often involves a tradeoff between costs in different stages of the supply chain, such as factory inventory, transportation, warehousing, manufacturing, etc. The economic optimization of supply chains was the first to be studied. In this chapter, we will discuss some commonly used economic impact quantification methods for business processes in general, while paying special attention to supply chain management.

1 Quantification of Economic Impact in a Business Context and Supply Chains

Since there is no universally accepted standard for the measures that comprise each of the three TBL categories, different economic measures have been applied to quantify the economic impact in literature. In general, economic performance indicators encompass the economic bottom line and the flow of money. A variety of economic measures are used depending on the business context (businesses or non-profits), different projects or policies (infrastructure investment or educational programs), or different geographic boundaries (a city, region, or country). The following subset is only an example (Slaper 2011): personal income, cost of

D. Inghels, *Introduction to Modeling Sustainable Development in Business Processes*, https://doi.org/10.1007/978-3-030-58422-1_4

Table 4.1 Indicators used for economic sustainability of supply chains according Mujkić et al. (2018)

Sustainability	Decision level	Indicators	Number of times used in scientific papers 2000-2017
Economic	Strategic	Profit	7
		Cash flow	6
		Delivery lead time	5
		Customer satisfaction	2
		Trade level	1
		Budget variance	1
	Tactical	Total costs	32
		Capacity utilization	3
		Production effectiveness	1
	Operational	Demand uncertainty	5
		Product quality	1
		Percentage of defects	1

underemployment, establishment churn, establishment size, job growth, employment by sector, revenue, etc.

In this chapter, we will pay special attention to the quantification of the economic impact measurement in supply chain management as it is an example of a valuable business process. Mujkić et al. (2018) investigated which economic performance measurement indicators for supply chain optimization were used in scientific papers between 2000 and 2017. The majority of papers used cost minimization as an objective function; it was commonly presented as a single objective function and included various cost types. This is also the conclusion of Seuring and Goldbach (2002) and Seuring and Müller (2008). By contrast, profit and cash flow maximization were used significantly less often. However, they are equally important from a general business context perspective.

The popularity of economic sustainability indicators used in supply chain management also differs depending on the decision level that is studied (Mujkić et al. 2018). Cost is most prevalent in economic studies on tactical issues, profit on strategic issues, and demand uncertainty on operational issues as listed in Table 4.1.

The economic indicators depicted in Table 4.1 are categorized based on decision impact; they are divided into three levels (Tajbakhsh and Hassini 2015):

- Strategic decisions in supply chains deal with the long-term impact of supply chains (e.g., network structure, facility location, supplier selection).
- Tactical decisions in supply chains deal with production, logistics, and suppliers.
- Operational decisions in supply chains focus on materials, energy flow and operation, and fulfilling demand.

2 Cost, Profit and Cash Flow

Since cost, profit, and cash flow are the most frequently used economic objectives to be optimized within supply chains, we briefly discuss their definition.

2.1 Cost

Supply chain cost encompasses a variety of costs created by material and information flows throughout the supply chain and between the supply chain partners. It involves costs for inventory carrying, handling and storage, inspection and testing, packaging and forwarding, duties and taxes, labor and overhead, manufacturing, and raw materials, among other things.

2.2 Profit

Profit or net cash flow is revenue minus costs. Profit is defined as "*the surplus remaining after total costs are deducted from total revenue, and the basis on which tax is computed and dividend is paid. It is the best-known measure of success in an enterprise. Profit is reflected in reduction in liabilities, increase in assets, and/or increase in owners' equity. It furnishes resources for investing in future operations, and its absence may result in the extinction of a company. As an indicator of comparative performance, however, it is less valuable than return on investment (ROI). Also called earnings, gain, or income*" (Business Dictionary 2020).

The profit objective for the Green Waste Valorization case expressed in Z_6 is an example of how to derive an objective function for profit. In Table 1.2, the revenue and cost for compost and the revenue and cost of the separate sales of wooden cuttings are depicted. The profit for compost and wooden cuttings is the revenue minus the costs to produce these products.

In supply chains, the term *supply chain profit* is commonly used to denote the profit that is realized over the supply chain.

Supply chain surplus is the value added by the supply chain function of an organization. It is calculated as the revenue generated from a customer minus the total cost incurred to produce and deliver the product.

2.3 Cash Flow

Cash flow is defined as "*incomings and outgoings of cash, representing the operating activities of an organization. In accounting, cash flow is the difference in amount of cash available at the beginning of a period (opening balance) and the amount at the end of that period (closing balance). It is called positive if the closing balance is higher than the opening balance, otherwise called negative. Cash flow is increased by (1) selling more goods or services, (2) selling an asset, (3) reducing costs, (4) increasing the selling price, (5) collecting faster, (6) paying slower, (7) bringing in more equity, or (8) taking a loan. The level of cash flow is not necessarily a good measure of performance, and vice versa: high levels of cash flow do not necessarily mean high or even any profit; and high levels of profit do not automatically translate into high or even positive cash flow*" (Business Dictionary 2020).

Cash flow is calculated to evaluate the expenses and (future) benefits of an investment or project, among other things. The *net cash flow* is the arithmetic sum of the receipts (+) and the disbursements (−) that occur at the same point in time.

To facilitate describing investment in cash flows, while taking annual compounding interest and annual payments into account, the following notation will be adopted:

i: the annual interest rate
n: the number of annual interest periods
P: a present principal sum
A: a single payment, in a series of n equal payments, made at the end of each annual interest period
F: a future sum (cash flow), n annual interest periods hence

In engineering economy studies, disbursements made to implement an alternative are considered to take place at the beginning of the period embraced by the alternative (Thuesen and Fabrycky 1993). Receipts and disbursements occurring during the life of the alternative are usually assumed to occur at the end of the year or interest period in which they occur. The "year-end" convention is adopted for describing cash flows over time and for developing applicable cash flow diagrams.

The derivation and use of interest factors and annual payments involve four important points in time (Thuesen and Fabrycky 1993):

1. The end of 1 year is the beginning of the next year
2. P is at the beginning of a year at a time regarded as being the present
3. F is at the end of the nth year from a time regarded as being the present
4. An A occurs at the end of each year of the period under consideration

We will now discuss some commonly used compound amount factors.

2.4 Single-Payment Compound-Amount Factor

Suppose that P is invested now and earns an annual interest i per year. The corresponding cash diagram is shown in Fig. 4.1. This transaction does not provide any payments until the investment is terminated.

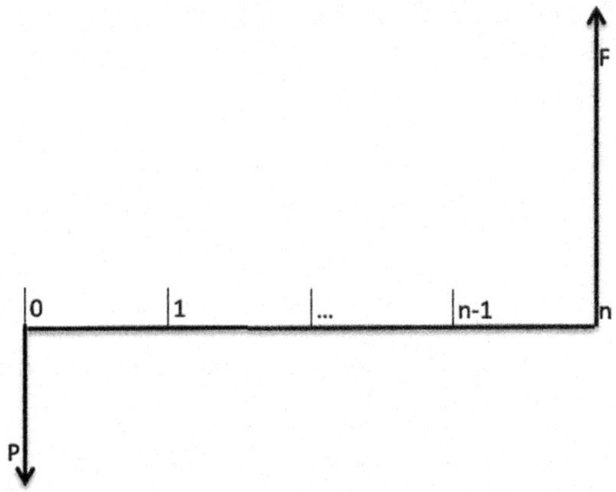

Fig. 4.1 Present and future value for a Single-Payment Compound-Amount

Table 4.2 Compound amount at the end of each year

Year	Amount at the start of the year	Interest earned during the year	Compound amount at the end of the year	
1	P	Pi	$P + Pi$	$= P(1 + i)^1$
2	$P(1 + i)$	$P(1 + i)i$	$P(1 + i) + P(1 + i)i$	$= P(1 + i)^2$
3	$P(1 + i)^2$	$P(1 + i)^2 i$	$P(1 + i)^2 + P$ $(1 + i)^2 i$	$= P(1 + i)^3$
n	$P(1 + i)^{n-1}$	$P(1 + i)^{n-1} i$	$P(1 + i)^{n-1} + P$ $(1 + i)^{n-1} i$	$= P$ $(1 + i)^n = F$

The amount at the beginning of the year is the amount at the beginning of the previous year plus the interest earned during the year as shown in Table 4.2. In general n years after the investment of P, the *future amount F* of a *present principal amount P* with an *interest rate i* is expressed in (4.1).

$$F = P \cdot (1 + i)^n \tag{4.1}$$

The resulting factor, $(1 + i)^n$, is known as the *Single-Payment Compound-Amount factor.*

The relation between the *present worth P* and the *future amount F* taking interest rate i into account over n years is written as:

$$P = F \cdot \left[\frac{1}{(1 + i)^n}\right] \tag{4.2}$$

The factor between brackets in (4.2) is known as the *Single-Payment Present-Worth* factor.

Let's use the Green Waste Valorization case as an example. In year 0, $P =$ €200,000 is invested in additional equipment to retrieve wooden cuttings in the pre-treatment section. At an interest rate of 7% per year, the compound amount at the end of the fifth year (i.e., the total amount to be paid back if the money is borrowed for 5 years) will be:

$$F = 200{,}000 \cdot (1 + 0.07)^5 = €280{,}510 \tag{4.3}$$

2.5 *Equal-Payment-Series Compound-Amount Factor*

Sometimes it is necessary to find the single future value that would accumulate from a series of equal payments with value A occurring at the end of n succeeding interest periods using an annual interest rate i. The series of cash flows are depicted in Fig. 4.2. The sum of the compound amounts of several payments equal to the future amount F is expressed in (4.4).

$$F = A \cdot \left[\frac{(1 + i)^n - 1}{i}\right] \tag{4.4}$$

The resulting factor between brackets in (4.4) is known as the *Equal-Payment-Series Compound-Amount factor*.

The reverse calculation delivers the *Equal-Payment-Series Sinking-Fund factor*, expressed between brackets in (4.5)

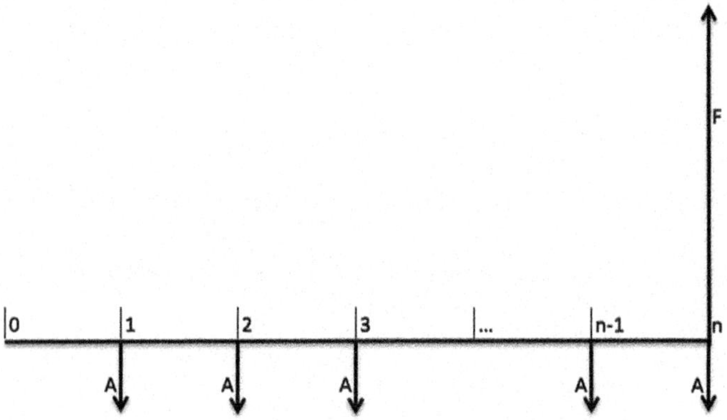

Fig. 4.2 Equal annual payments and future value for an Equal-Payment-Series Compound-Amount

$$A = F \cdot \left[\frac{i}{(1+i)^n - 1} \right] \tag{4.5}$$

Let's take the Green Waste Valorization case as an example once again. The future value F = €280,510 was calculated in (4.3) and an annual interest rate $i = 7\%$ applies. Suppose this future amount F will be paid back in equal amounts over $n = 5$ years. In that case, the single yearly payment A using Eq. (4.5) will be:

$$A = 280,510 \cdot \left[\frac{0.07}{1.07^5 - 1} \right] = 48,778 \ [\text{€/year}] \tag{4.6}$$

The last scenario that we will discuss is the evaluation of replacement alternatives. In this case, it is common to call an existing asset being considered for replacement the *defender*, and the asset proposed as a replacement the *challenger*.

Taking the Green Waste Valorization case as an example, let's suppose that 20,000 tons of biomass consisting of 20% wooden cuttings are composted with a compost yield of 35% each year. Using the figures from Table 1.2 in Chap. 1, the profit from selling compost is €5/ton. Therefore, the annual income from selling the compost is:

$$A = 20,000 \cdot 0.35 \cdot 5 = 35,000 \ [\text{€/year}] \tag{4.7}$$

By sieving the compost, the wooden cuttings can be retrieved during an extra pre-treatment step. However, this requires a €200,000 investment in additional equipment. The annual income from selling compost will drop by 20% if the wooden cuttings are removed. The annual profit related to the sale of the compost, A_c, will also drop by 20%. In other words: $A_c = 35,000 \cdot 0.8 = 28,000$ [€/year]. The profit from selling wooden cuttings is €9/ton. Therefore, the annual profit from selling wooden cuttings, A_w, will become $A_w = 20,000 \cdot 0.2 \cdot 9 = 36,000$ [€/year]. The investment will take place if the net profit each year is assumed to be positive, including the investment with P = €200,000 will be paid back in $n = 5$ years with annual payments at an interest rate of $i = 7\%$. We know from (4.6) that this is equal to an annual sum of 48,778 [€/year] to be paid back.

In this case, the current or defender situation is A = 35,000 [€/year] and represents the complete allocation of green waste to composting.

The challenger situation is the one with the investment of P = €200,000, paid back over $n = 5$ years, with equal annual payments, and yielding an annual profit of $A_c + A_w$ = 28,000 + 36,000 = 64,000 [€/year] which is larger than the profit A = 35,000 [€/year] of the defender situation.

The challenger situation yields an annual positive net profit of 64,000–48,778 = 15,222 [€/year]. Consequently, the investment will take place.

The output of these calculations can be taken into account when formulating the objective function describing the economic impact that needs to be optimized in the problem under study.

3 Internalizing External Costs

The three pillars of sustainability are generally expressed in different units. This is not a problem when formulating a MOOP. Another approach that is used is monetizing all the dimensions of the triple bottom line, including environmental damage and societal costs such as noise pollution or costs of accidents on roads. Taking the environmental and societal life-cycle impact into account is also known as internalizing the external costs or life-cycle cost (LCC) or lifetime cost. The life-cycle cost of an item can be defined as the sum of all funds spent in support of the item from its conception and fabrication through its operation to the end of its useful life (Woodward 1997).

The advantage of this approach is obvious. By expressing the triple bottom line in monetary terms, the optimization of a sustainability problem is reduced to a SOOP expressed in monetary terms. However, estimating environmental and societal costs is highly dependent on assumptions and data collection. Therefore, there is always some uncertainty associated with life-cycle costs. For example, consider Table 4.3 discussed in Inghels et al. (2016c), expressing the external costs per transportation mode in the EU and Table 4.4 expressing the same external costs in a different study.

Table 4.3 External costs per transport mode applicable for the EU (EC 2008) according Inghels et al. (2016c)

Average external cost [€/ 10^3 tkm]	Truck	Barge	Train
Accidents	5.4	0.0	1.5
Noise	2.1	0.0	3.5
Air pollution and climate	8.7	3.0	4.3
Congestion	5.5	NA	0.2
Infrastructure	2.5	1.0	2.9
Total	**24.10**	**5.0**	**12.3**

Table 4.4 External costs per transport mode (in Euro/1000 tkm) for EU-27 (table 3.2 Panteia 2013) according Inghels et al. (2016c)

2011	Climate change costs	Air pollution costs	Total average external costs [€/ 10^3 tkm]
Road (Truck)	€ 6.95	€ 7.00	€ 13.95
IWT (Barge)	€ 3.06	€ 10.47	€ 13.53

Chapter 5
Quantification of the Environmental Impact

In this chapter, we focus on quantifying the environmental pillar of the People-Planet-Profit triple bottom line by introducing life cycle assessment (LCA). Life cycle assessment is rooted in life cycle thinking, which approaches products, services, and production systems holistically in terms of impact on the environment. Life cycle thinking advocates taking into account the environmental impact of every stage in the life cycle, from the extraction of raw materials, material processing, transportation, distribution, consumption, reuse/recycling, to disposal. Life cycle assessment is a well-established analytical method for quantifying the environmental impact of a product, service, or production process. This method is traditionally used to study four types of problems: (i) assessment of individual products to understand their environmental impact, (ii) comparison of process paths in the production of substitutable products or processes, (iii) comparison of alternatives for delivering a given function (Jacquemin et al. 2012), and (iv) to analyze the phases of the product or service life cycle that have a more considerable environmental impact – also known as "hotspots" (Piekarski et al. 2013). Several industries, companies, and associations are actively developing LCA approaches. The European Union (EU) is one of the leading global regions advocating life cycle thinking and assessment to ensure the identification of the best environmental outcome.

1 Introduction

1.1 The Generic Life Cycle of a Product

A general overview of a product life cycle is depicted in Fig. 5.1. The material life cycle starts with the extraction of raw materials, selected in the design phase, that will be used in the final product under study. Before these raw materials can be used in manufacturing and assembly, they have to be transformed into usable materials

D. Inghels, *Introduction to Modeling Sustainable Development in Business Processes*, https://doi.org/10.1007/978-3-030-58422-1_5

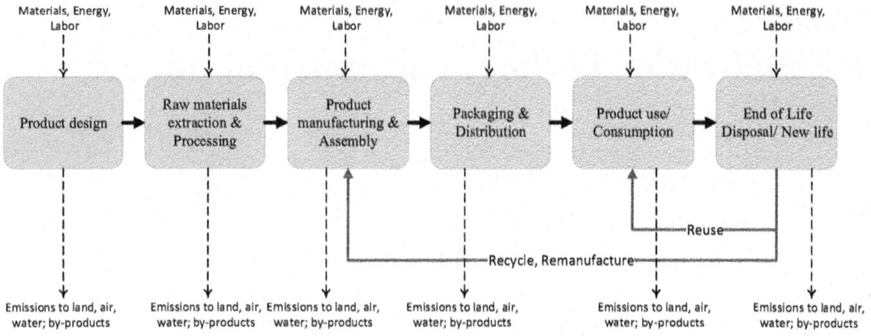

Fig. 5.1 Generic life cycle of a product. (Based on Life Cycle Initiative 2020)

first. Once a product is manufactured or assembled and subsequently packaged and distributed, the usage or consumption phase starts. During the usage phase, service activities can take place to repair or maintain the product. After a useful life, the end-of-life recovery process is initiated, or the product is discarded. Numerous material recovery and waste management options are possible in this stage: materials can be reused, remanufactured, or recycled and used as secondary material for making new products.

Each step in the life cycle is, in general, characterized by the consumption of materials, energy, and commonly available substances such as water or air together with labor (Life Cycle Initiative 2020). The output of each intermediate step is the input of a subsequent step in the life cycle. Additional inputs for each step in the life cycle can come from the reuse, remanufacturing, and recycling of goods. Moreover, water, air, or land emissions may be generated in each step together with by-products. By-products are secondary products derived from a production process that can be marketable or considered as waste. If recycling or remanufacturing is involved in the life cycle of a product, this is called closed-loop manufacturing. Closed-loop manufacturing is at the basis of the circular economy, which focuses on the elimination of waste and the minimization of energy and raw material consumption, thereby minimizing the pollution associated with products and services. If products are not recycled or remanufactured, we call this the "take-make-dump" approach to open-loop manufacturing, which is characteristic of the traditional economy. The depletion of raw materials and common substances, together with the creation of waste and emissions, have an impact on the environment, which is affected by life cycle impact categories such as global warming potential, ozone depletion potential, etc. These will be discussed later in this chapter.

A substantial amount of energy and greenhouse gas emissions can often be saved using closed-loop recycling and remanufacturing to make new products. Recycled materials are also known as secondary raw materials. Take aluminum as an example: using secondary aluminum can save up to 95% of energy over the entire life cycle compared to primary aluminum from bauxite, the principal raw material for making aluminum (BIR 2008). That is because producing primary aluminum is an energy-intensive process that involves electrolysis. Aluminum is increasingly used in cars to reduce their environmental footprint. However, more than double the amount of CO_2 gasses are emitted when manufacturing semi-finished aluminum compared to making the same quantity of steel. Nevertheless, using lightweight aluminum instead of steel realizes net CO_2 emission savings over the entire life cycle of a passenger car with an internal combustion engine due to reduced fuel consumption. Using an increased amount of recycled aluminum in the production of semi-finished aluminum will result in a further drop in CO_2 gas emissions when used to manufacture the same automotive parts. During recycling, a downgrading of the recycled aluminum can take place when the recycled aluminum is not kept separate per alloy. Downgrading ultimately reduces the number of applications for which the recycled aluminum might be suitable. Note that the energy and CO_2 emissions savings from using secondary raw materials are not only applicable to the aluminum example; they are also applicable to lots of other materials, as depicted in Table 5.1 (BIR 2008).

The same study also revealed the associated CO_2 emissions and the avoidance of these emissions while using secondary instead of primary raw materials (Table 5.2).

Table 5.1 Energy consumption per 100,000 tons of production and savings by using secondary instead of primary raw materials

Material	Primary [TJ/100,000 ton]	Secondary [TJ/100,000 ton]	Energy savings using secondary versus primary [%]
Aluminum	4700	240	95
Copper	1690	630	63
Ferrous	1400	1170	16
Lead	1000	13	99
Nickel	2064	186	91
Tin	1820	20	99
Zinc	2400	1800	25
Paper	3520	1880	47

Source: BIR (2008)

Table 5.2 Carbon footprint and savings expressed in kilotonnes of CO_2 ($ktCO_2$)/100,000 tonnes and CO_2 savings using secondary instead of primary materials

Material	Primary [ktonCO_2/ 100,000 ton]	Secondary [ktonCO_2/ 100,000 ton]	Savings CO_2 emissions [%]
Aluminum	383	29	92
Copper	125	44	65
Ferrous	167	70	97
Lead	163	2	99
Nickel	212	22	90
Tin	218	3	99
Zinc	236	56	76
Paper	0.17	0.14	18

Source: BIR (2008)

1.2 The Purpose of Life Cycle Assessment

The life cycle assessment of a product collects all the inputs and outputs over every stage in the life cycle process of this product, enabling the calculation of the total environmental impact. The total environmental impact is expressed in different life cycle impact categories, which will be discussed later in this chapter.[1]

The total environmental impact assessment of a product makes it transparent where the highest environmental impact occurs in the life cycle for both the producer and the consumer. It allows producers and consumers to take actions to lower the environmental burden associated with products, as shown in the Levi Strauss example below.

According to Levi Strauss & Co. (2015), a pair of Levi's jeans consume, on average, 3781 liters of water during its life cycle. The water is mainly used to grow the cotton and manufacture the pair of jeans. During the usage phase, an average of 860 liters of water is consumed to wash the jeans according to Levi's. The usage phase accounts for about 23% of the total water used over the lifetime of a pair of Levi's jeans while growing the cotton accounts for about 68% of the total water consumption. Consumers can impact the environmental footprint of a pair of jeans in terms of water consumption by washing the jeans less frequently and by using cold water.

The global warming impact of a pair of Levi's jeans is 33.4 kg CO_{2e} (see Fig. 5.2). This figure is impacted by the usage phase. Consumers have a high impact on the associated CO_{2e} emissions in this phase because they choose how many times the jeans are worn before they are washed. The subsequent drying process has an even higher impact: washing 1 kg of clothing at 40 °C consumes 0.21 kWh of energy, but tumble drying it requires three to four times as much energy. By

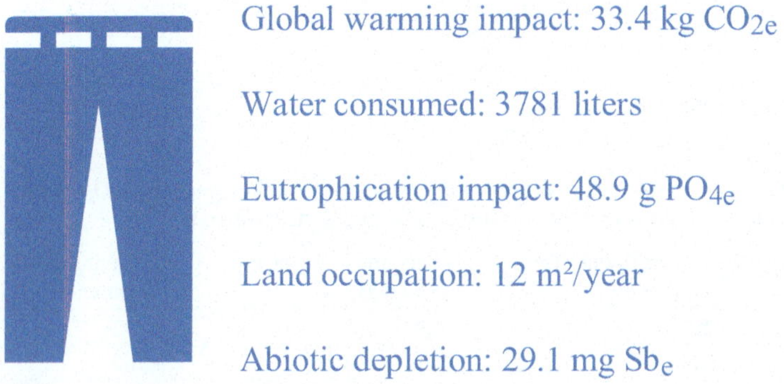

Fig. 5.2 Lifecycle impact of a pair of Levi's® 501® jeans. (Source data: http://levistrauss.com/wp-content/uploads/2015/03/Full-LCA-Results-Deck-FINAL.pdf)

[1]see also http://www.lifecycleinitiative.org/applying-lca/lcia-cf/

replacing tumble drying with line drying, consumers can positively impact global warming. Tumble drying is used more than 80% of the time in the United States compared to 20% in Germany and Sweden and 12% in Poland (Gwozdz et al. 2017).

2 Commonly Used Terminology

In this section, we list the most important terms used in environmental claims, as defined in the EN 14021 standard (2016) on environmental labels and declarations, and some important terms used in life cycle assessment, as defined in the EN 14040 standard (2006). These terms are used in the remainder of this chapter or are of importance in this field.

Allocation: partitioning the input or output flows of a process or a product system between the product system under study and one or more other product systems

Comparative assertion: environmental claim regarding the superiority or equivalence of one product versus a competing product that performs the same function

Cut-off criteria: specification of the amount of material or energy flow or the level of environmental significance associated with unit processes or product system to be excluded from a study

Critical review: process intended to ensure consistency between a life cycle assessment and the principles and requirements of the International Standards on life cycle assessment

Environmental aspect: element of an organization's activities, products or services that can interact with the environment

Environmental impact: any change to the environment, whether adverse or beneficial, wholly or partially resulting from an organization's activities or products

Impact category: class representing environmental issues of concern to which life cycle inventory analysis results may be assigned

Interested party: individual or group concerned with or affected by the environmental performance of a product system, or by the results of the life cycle assessment

Life cycle: consecutive and interlinked stages of a product system, from raw material acquisition or generation of natural resources to final disposal

Life cycle assessment (LCA): compilation and evaluation of the inputs, outputs, and potential environmental impacts of a product system throughout its life cycle

Life cycle inventory analysis (LCI): phase of life cycle assessment involving the compilation and quantification of inputs and outputs for a product throughout its life cycle

Life cycle impact assessment (LCIA): phase of life cycle assessment aimed at understanding and evaluating the magnitude and significance of the potential environmental impacts for a product system throughout the life cycle of the product

Life cycle interpretation: phase of life cycle assessment in which the findings of either the inventory analysis or the impact assessment, or both, are evaluated in relation to the defined goal and scope in order to reach conclusions and recommendations

*Product***:** any goods or service

Reference flow: measure of the outputs from processes in a given product system required to fulfill the function expressed by the functional unit

Waste: anything for which the generator or holder has no further use and which is discarded or is released into the environment

Recycled content: proportion, by mass, of recycled material in a product or packaging. Only pre-consumer and post-consumer materials are considered recycled content.

Pre-consumer material: Material diverted from the waste stream during the manufacturing process. Re-utilization of materials such as rework, regrind, or scrap generated in a process and capable of being reclaimed within the same process that generated it is excluded.

Post-consumer material: Material generated by households or by commercial, industrial, and institutional facilities in their role as end-users of a product that can no longer be used for its intended purpose. This includes returns of material from the distribution chain.

Recycled material: material that has been reprocessed from recovered [reclaimed] material by means of a manufacturing process and made into a final product or a component for incorporation into a product

Recovered (reclaimed) material: material that would have otherwise been disposed of as waste or used for energy recovery, but has instead been collected and recovered [reclaimed] as a material input, in lieu of new primary material, for a recycling or manufacturing process

System boundary: set of criteria specifying which unit processes are part of a product system

3 The Four-Stage Approach of an LCA According to ISO 14040+44

The ISO 14040 (2006) standard on Environmental management – Life cycle assessment, Principles and framework – and the ISO 14044 standard on Environmental management – Life cycle assessment, Requirements and guidelines – prescribe a clear and structured approach on how to conduct an LCA. Many life cycle assessments are conducted according to the four-stage approach advocated in the ISO 14040 and ISO 14044 standards (see Fig. 5.3):

1. Goal and Scope Definition
2. Inventory Analysis
3. Impact Assessment
4. Interpretation

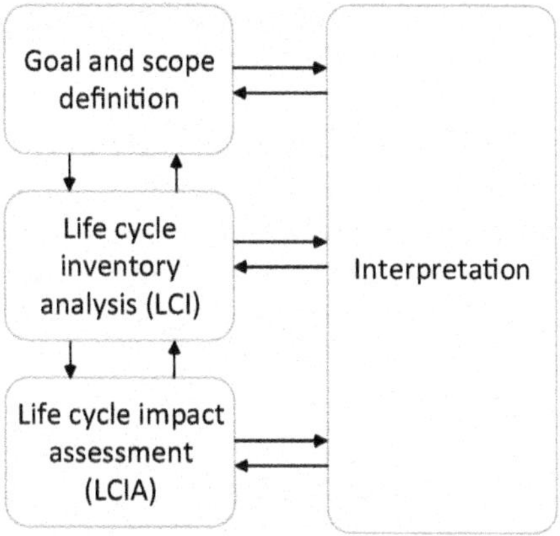

Fig. 5.3 The four stages of an LCA according to ISO 14040 (2006)

In the next sections, we briefly explain the four stages as prescribed by the ISO 14040 (2006) standard.

3.1 *Goal and Scope Definition*

In the *goal* definition of an LCA, the objectives and applications are defined. It states why the LCA study is carried out, for whom the results of the study are intended, called the *intended audience*, and what can be done with the results of the LCA study. Furthermore, it is important to mention whether the results are intended to be used in comparative assertions intended to be disclosed to the public. Comparative assertions require a critical review by an internal or external expert who is familiar with the requirements of LCA and who should have the appropriate scientific and technical expertise. The critical review is introduced since the outcome of the comparative LCA might affect interested parties that are external to the LCA.

The *scope* definition is added to ensure that the breadth, depth, and detail of the study are compatible and sufficient to address the stated goal. The scope of the LCA study addresses issues related to the function of the product under study or the functions of the systems under study in case of a comparative LCA and its system boundaries and the data categories that are used. In the next subsections, we discuss the most important issues related to the scope definition.

3.1.1 Functional Unit

LCA models the life cycle of a product as its product system, which performs one or more defined functions. The essential property of a product system is characterized by its function(s).

The function of a product is quantified by the *functional unit.* The functional unit defines what is being studied. As an example, the functional unit for a pair of Levi's 501 jeans is, according to Levi Strauss & CO (2015), a "*Levi's® 501® jeans & other core LS & Co. products*". The Levi's® 501® product attributes studied are: five fabrics, eight finishes (low to high complexity; highest volume), and the 2012 production year. Another example of a functional unit is the function of a beverage container, which can be a bottle in glass or plastic or a can. The functional unit of such a beverage container can be defined as the stored volume in the beverage container (e.g. 333 ml).

3.1.2 Reference Flow

Next to the functional unit, the *reference flow* has to be defined in the scope of the LCA. The reference flow indicates the quantity of products (components and materials) needed to fulfill the intended function as defined by the functional unit. As an example, the LCA by the EAA (2013) on the update of its various European LCI datasets related to aluminum processes states that the reference flow used is "*the production of 1 ton of ingot from primary aluminum, i.e., from bauxite mining up to the sawn aluminum ingot ready for delivery*".

3.1.3 System Boundary

The scope definition should also include the *system boundary,* which defines the unit processes to be included in the system. The choice of unit processes to be included in the LCA model depends on the goal and scope defined in the LCA study. According to ISO 14040 (2006), ideally, the product system should be modeled in such a manner that inputs and outputs at its boundary are elementary flows. Elementary flows include the use of resources that are associated with the system as well as emissions into the air, water, and land.

When defining the system boundary for a specific LCA, several life cycle stages, unit processes, and flows should be taken into consideration. Figure 5.4 is an example of the system boundaries of the various LCI datasets that are the subject of the LCA related to aluminum processes (EAA 2013). The generic life cycle stages depicted in Fig. 5.1 reappear in the system boundaries of Fig. 5.4. Raw material extraction and processing is the subject of the first LCI dataset named "Primary dataset (A)". Semi-production is another LCI dataset; it is indicated next to the use of products made out of aluminum and the collection and sorting at the end of life.

System boundaries are commonly referred to. They include the following (see Fig. 5.5):

- *Cradle to Grave*: includes the material and energy production chain and all processes from raw material extraction to end-of-life treatment, including the production, transportation, and use phase.

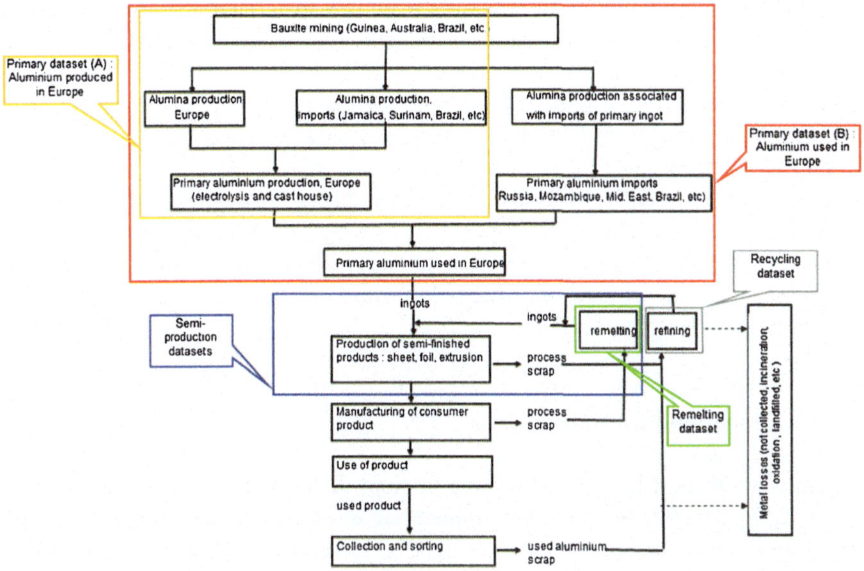

Fig. 5.4 Overview of the system boundaries for the LCI datasets related to aluminum processes. (Source: EAA 2013)

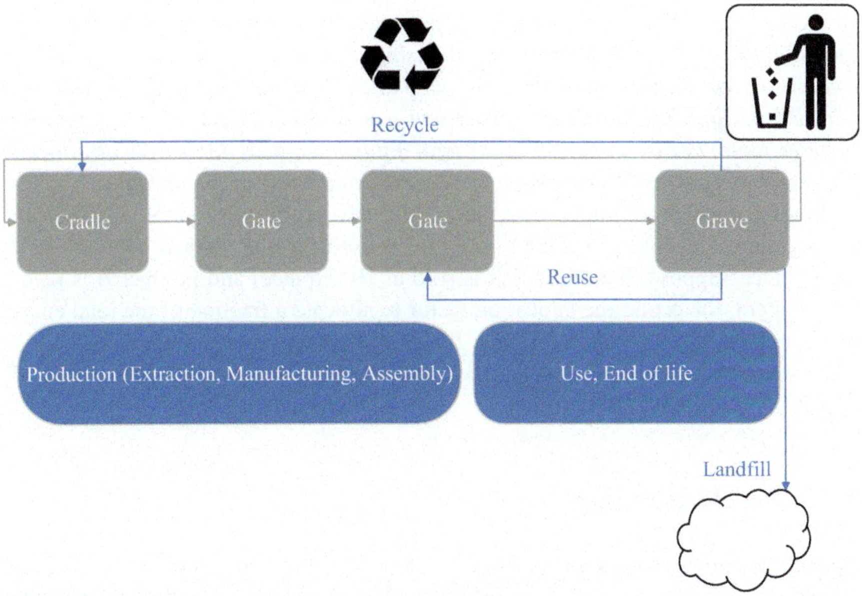

Fig. 5.5 Life cycle stages and system boundaries

- *Cradle to Gate*: includes all processes from raw material extraction through the production phase (i.e., the factory gate). This type of boundary is often used to determine the environmental impact of the production of a product.
- *Gate to Grave*: this is used to determine the environmental impact of a product once it leaves the factory, including all post-production processes (i.e., usage and end-of-life treatment).
- *Gate to Gate*: only covers the production phase processes. This is used to determine the environmental impact of a single production step or process.
- *Cradle to Cradle*: extends the Cradle to Grave boundaries with the premise that all used materials at the end-of-life will be used in a new product without quality loss or wasted by-products.

3.1.4 Allocation Procedures

The allocation must be defined if there is more than one input or output in a unit process. Suppose that two different products are assembled in the same factory using a common power supply. Since there will be no separate recording of the power used to assemble one of the two products, this power supply must be allocated when an LCA is made for one of these products. Allocation requires an allocation factor; this can be a physical quantity (e.g., mass, volume, or energy content) or an economic value (e.g., sales price or net cost).

According to ISO 14041, the procedure for non-physical relationship allocation is the following: "*Where physical relationship (i.e. kg, m^2, m^3, etc.) cannot be established or used as the basis for allocation, the inputs should be allocated between the products and the functions in a way that reflects other relationships between them. For example, environmental input and output data might be allocated between co-products in proportion to the economic value of the products.*"

In the case of a common power supply for two products assembled in the same factory, the economic allocation of the power used for one product could be in the sales price. Suppose that product A is sold at 10 [€/piece] and product B is sold at 15 [€/piece], the economic allocation factor to allocate a fraction of the total energy input to product A is 10/(10 + 15). If the total energy consumption for a given year was 250,000 [kWh], the energy allocation to product A is (10/25)· 250,000 = 100,000 [kWh].

3.1.5 Data Requirements

The life cycle inventory analysis (LCI) phase is the second phase of LCA. It encompasses an inventory of input/output data with regard to the system being studied. The data collection requirements for the LCI should be described in the goal and scope phase.

An example of such a data collection description can be found in EEA (2013): "*The present life cycle inventory data for aluminum is derived from various industry* *surveys covering the year 2010. The various European plants participating in the survey delivered absolute figures of process inputs/outputs for the whole year 2010 (tons, GJ, m^3, etc.). After aggregation, these input and output data were used to calculate European averages. These European averages were then integrated within specific LCI models in order to generate the corresponding LCI datasets.*"

3.1.6 Cut-Off Criteria

Cut-off criteria to the inputs and outputs of the LCA data are commonly applied to deal with the tradeoff between resources needed to gather all the data and the significance of using these data on the output of the LCA. ISO 14040 (2006) states that: "*The cut-off criteria used within a study should be clearly understood and described.*"

Let us look at the example found in EEA (2013) once again: "*Input and output data have been collected through detailed questionnaires which have been developed and refined from the first surveys organized in 1994–1996. In practice, this means that, at least, all material flows going into the aluminum processes (inputs) higher than 1% of the total mass flow (t) or higher than 1% of the total primary energy input (MJ) are part of the system and modelled in order to calculate elementary flows. All material flows leaving the product system (outputs) accounting for more than 1% of the total mass flow is part of the system. All available inputs and outputs, even below the 1% threshold, have been considered for the LCI calculation. For hazardous and toxic materials and substances the cut-off rules do not apply.*"

3.1.7 Selected Impact Categories and Methodology of Impact Assessment

The scope definition phase must mention the methodology of the life cycle impact assessment and the selected impact categories. An example of a frequently used life cycle impact assessment (LCIA) methodology is the scientific CML process (see http://www.cml.leiden.edu). CML uses the following impact categories: the Global Warming Potential (GWP) expressed in [$kgCO_{2\text{-eq}}$/kg], the stratospheric ozone layer depletion (ODP) expressed in [kg $CFC11_{\text{-eq}}$/kg], the summer smog creation potential (POCP), the acidification potential of soils and water bodies (AP) expressed in [kg $SO_{2\text{-eq}}$/kg], the nitrification potential of soils and water bodies (NP), and the depletion of abiotic resources (ADP) expressed in [1/year]. Other internationally accepted impact assessment methodologies are eco-indicator 99, IMPACT 2002+, ReCiPe, TRACI I, and EDIP.

Table 5.3 is an example of the impact categories, the units used, and the impact assessment methodology used in the LCA on the European LCI datasets related to aluminum processes (EAA 2013):

Table 5.3 The selected impact categories used in the LCA on the European LCI datasets (EAA 2013)

Impact categories	Unit	Methodology
Depletion of Abiotic Resources (ADP)	[kg Sb-Equiv.]	CML2001- Nov 2010
Acidification Potential (AP)	[kg SO_2-Equiv.]	CML2001- Nov 2010
Eutrophication Potential (EP)	[kg Phosphate-Equiv.]	CML2001- Nov 2010
Greenhouse Gas emission (GWP 100 years)	[kg CO_2-Equiv.]	CML2001- Nov 2010
Ozone Layer Depletion Potential (ODP, steady state)	[kg R11-Equiv.]	CML2001- Nov 2010
Photo-oxidant Creation Potential (POCP)	[kg Ethene-Equiv.]	CML2001- Nov 2010
Total Primary energy	[MJ]	net cal. value
Primary energy from renewable raw materials	[MJ]	net cal. value
Primary energy from non-renewable resources	[MJ]	net cal. value

3.1.8 Other Elements Belonging to the Goal and Scope Phase

The following elements can also be added to the most important elements of the goal and scope phase:

- Assumptions: in an LCA, one may rely on some assumptions regarding the choice of elements of the physical system being modeled or the selection, modeling, and evaluation of impact categories. When comparing the results of different LCA studies, the underlying assumptions must be equivalent. For transparency reasons, assumptions must be clearly described and reported.
- Limitations of the LCA: if constraints are placed on excluding some input or outputs on unit processes, or when data gaps are encountered, one should document these limitations in the LCA report.
- Initial data quality requirements: the quality of LCA results depends on the quality of the input data and the quality of the methodology or model that is used. The quality of the input data is equivalent to the degree of confidence in individual input data or the data set as a whole.
- Type of critical review, if any: according to ISO 14040 (2006), critical review is a process to verify whether an LCA has met the requirements for methodology, data, interpretation, and reporting and whether it is consistent with the principles. In the case of an LCA to support comparative assertions, a critical review by an external expert must be conducted since the conclusions might affect interested parties. The scope and type of critical review are defined in the scope phase of the LCA.
- Type and format of the report required for the study. The LCA report should address the different phases of the LCA study. The results and conclusions,

together with the data, methods, limitations, and assumptions applied in the study, should be reported in an adequate form to the intended audience.

3.2 *Life Cycle Inventory Analysis (LCI)*

According to ISO 14040 (2006), the life cycle inventory analysis (LCI) involves data collection and calculation procedures to quantify relevant inputs and outputs of a product system.

The LCI starts with the collection of data for each unit process within the system boundary. The following major data categories, which should be defined in the scope definition, can be considered (ISO 14040 2006):

- Energy inputs, raw material inputs, ancillary inputs, other physical inputs
- Products, co-products, and waste
- Emissions to air, discharges to water and soil
- Other environmental aspects

The LCI phase is, in general, a resource-intensive process that can comprise collected data together with data from validated sources. The data sources that are considered in the LCA must be documented in the scope of the LCA.

As example, the LCI of the unit process "bauxite mining" of Fig. 5.4 is depicted in Table 5.4 (EAA 2013). It encompasses the direct input and output data for the extraction and preparation of one ton of bauxite. The input and output data have been taken from the worldwide IAI survey based on the year 2010.

Table 5.4 Example of an LCI of the unit process "bauxite mining" (EAA 2013)

	Unit	Bauxite mining world (IAI) 2010
Inputs		
Raw materials		
Fresh water	m^3/t	0.5
Sea water	m^3/t	0.7
Fuels and electricity		
Heavy oil	kg/t	0.2
Diesel oil	kg/t	0.3
Electricity	kWh/t	0.9
Outputs		
Air emissions		
Carbon dioxide (CO_2)	kg/t	2
Particulates	kg/t	0.17
Water discharge		
Fresh water	m^3/t	0.05
Sea water	m^3/t	0.7

Once the data is collected, it must be validated. The data must be related to the defined unit processes and the reference flow of the functional unit.

3.3 *Life Cycle Impact Assessment (LCIA)*

In the life cycle impact assessment phase, the LCI results gathered in the previous phase are converted into potential impacts on the environment. The selected impact categories and methodology of impact assessment, as defined in the scope of the LCA, are employed. This phase provides information for the life cycle interpretation phase.

The LCIA consists of two other mandatory elements (see Fig. 5.6): classification and characterization. The goal of classification is to link the LCI data categories or parameters (CO_2, methane, SO_x, NO_x, etc.) to their relevant life cycle impact categories (Global Warming Potential, Acidification Potential, etc.). Characterization consists of three steps (Wimmer et al. 2004):

1. Quantification of the environmental impact of an inventory parameter belonging to a given impact category
2. Repeat the first step for all inventory parameters belonging to the same impact category and sum them up
3. Repeat the first and second steps for all impact categories.

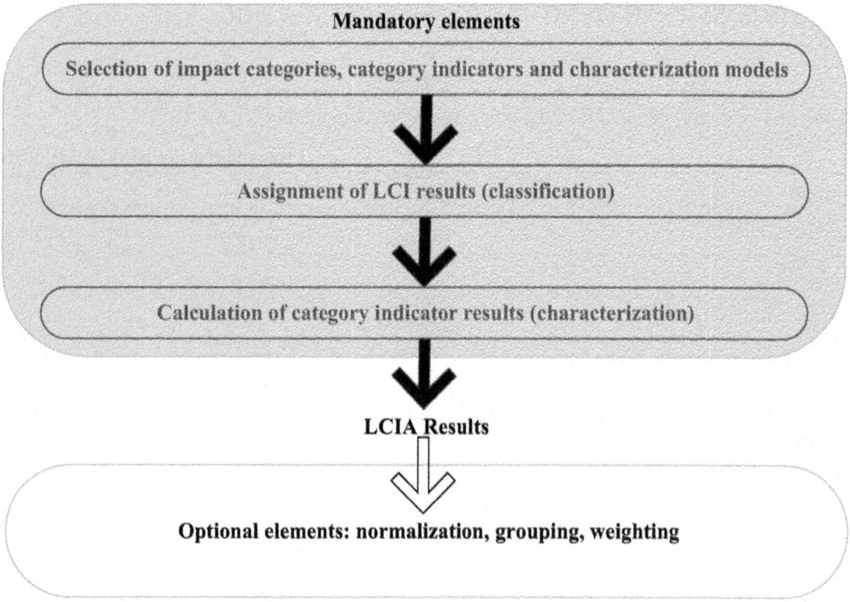

Fig. 5.6 Elements of the LCIA phase. (Source: ISO 14040 2006)

Suppose that the output of emissions of a product are CO_2, CH_4 (methane), and N_2O (nitrous oxide). All three different types of emissions have an impact on global warming that is expressed in [kg $CO_{2\text{-eq}}$/kg]. Characterization expresses the environmental impact of these emissions on the equivalent impact of the characterization factor unit of the environmental impact category. In this case, the global warming potential of the different emissions is converted into CO_2 equivalent emissions. The factors 25 for CH_4 and 298 for N_2O in the example below are called the characterization factors.

$$1\ \text{kg}\ CO_2 \Leftrightarrow 1\ \text{kg}\ CO_2 - \text{equivalent}$$

$$1\text{kg Methane}\ (CH_4) \Leftrightarrow 25\text{kg}\ CO_2 - \text{equivalent}$$

$$1\ \text{kg Nitrous oxide}\ (N_2O) \Leftrightarrow 298\ \text{kg}\ CO_2 - \text{equivalent}$$

3.4 *Interpretation*

The last phase of an LCA is the interpretation phase. In the interpretation phase, the findings from the inventory analysis and the impact assessment are considered together. In the case of LCI studies, only the findings of the inventory analysis are mentioned. The findings may be expressed as conclusions and recommendations for the intended audience, as defined in the goal and scope phase of the LCA. According to ISO 14040 (2006), the interpretation phase should deliver results that are consistent with the defined goal and scope and which reach conclusions, explain limitations, and provide recommendations.

An example of how LCI and LCIA data can be used in the interpretation phase is the creation of awareness that the environmental impact of a pair of Levi's jeans can be influenced during the usage phase by the washing and drying behavior of the consumer.

4 LCA Applied to the Green Waste Valorization Case Study

In this section, we briefly explain the LCA on green waste carried out by SenterNovem (2008), which is the basis for the environmental impact assessment of the Green Waste Valorization study case discussed in Chap. 1.

4.1 Goal and Scope

Goal and Scope: The goal of this LCA is to determine and compare the environmental impact associated with the incineration and composting of green waste. The obtained information can be used to review the policy towards green waste management as input in the LAP2 in the Netherlands (LAP2 = "Landelijk Afvalbeheerplan 2" or the Rural Waste Management Plan). The results are made available to the public. In order to determine the sensitivity of the two green waste management valorization methods, a sensitivity analysis is carried out.

Functional unit: A batch of one ton of green waste with average composition in the Netherlands

System boundary: The process of green waste management is best evaluated on its own using a *gate-to-gate* approach since green waste is a natural product, unlike products that have a complete product life cycle (Grant 2003). Therefore, the Green Waste Valorization problem only considers the processes that deal with composting and incineration with energy valorization. The LCA does not take the collection and transportation of green waste into account since this is the same for both valorization options. The scope is to compare the valorization options of composting and incineration with energy recuperation. However, SenterNovem (2008) takes into account the environmental effects of substituting green waste for composting (i.e., peat replacement) and for energy valorization (i.e., the replacement of coal).

In order to conduct the LCA, the following information has been gathered:

- Composition of the green waste: the composition of the green waste taken into account is not the average composition of green waste in the Netherlands, but an approximation of the composition ready to be used in biomass power plants. These power plants are only interested in the parts of the green waste with low humidity. For all the different fractions, the Phyllis database is used. This guarantees that values pertaining to the Netherlands with respect to the composition and humidity of the fractions in the green waste (i.e., leaves, grass, and prunings) are used.
- The energy consumption of the processes included in the LCA
- The consumption of consumables such as chemicals, water, etc. used in the processes included in the LCA
- The emissions to soil and groundwater
- The emissions to air, soil, and groundwater that were avoided by using green waste as a source of input

4.2 Life Cycle Inventory

The LCI of Green Waste is depicted in Table 5.5:

The LCA assessment data originates from two reference installations that were used to analyze the environmental performance of composting and incineration with energy recuperation (Table 5.6).

The "Composteerinrichting Boeldershoek" is a composting plant located in Boeldershoek; it is a modern green waste composting facility. The throughput time of the composting process takes about 3 months; it requires a mixture of green waste, open-air composting, and the compost must be sieved.

The obtained compost can be used as a high or low value substitute: it is a high value substitute if it is used to replace chemical fertilizers. During the composting process, methane (CH_4) is emitted into the air, which has a global warming impact that is 25 times higher than CO_2 gas emissions. All the emissions to the soil are also taken into account.

The "BES verbrandingsinstallatie te Sittard" is a Waste-to-Energy plant in Sittard. It hosts a stand-alone incinerator that can incinerate green waste and allows for thermic wood treatment. During the incineration process, this plant generates power and heat, which is used in a local heating network. A denox cleaning of the flue gas (i.e., a waste or by-product of incineration) takes place. All emissions to the air are taken into account in the life cycle inventory. The effect of landfilling some remainders after incineration is taken into account, as is the avoidance of other combustibles for the generation of power and heat.

Table 5.5 composition of green waste to be used in biomass plants in the Netherlands by SenterNovem (2008)

			Different composition (for sensitivity analysis)		
Component	Unit	Standard composition	1	2	3
Wooden cuttings (prunings)	Weight% d.m.	60	60	50.8	60
Leaves	Weight% d.m	20	20	6.7	20
Grass	Weight% d.m	20	20	42.5	20
Ash	Weight% d.m	20	20	20	25
Humidity	Weight% ar	50	45	55	55
NCV (ar)		6.4	7.3	5.5	5.2

dm dry mass, *NCV* Net Caloric Value, *ar* relative atomic mass

Table 5.6 List of reference installations

Green waste valorization	Reference installation
Composting	Composteringsinrichting Boeldershoek
Incineration with energy recuperation	BES verbrandingsinstallatie te Sittard

4.3 Life Cycle Impact Assessment

The outcome of the LCIA of the Green Waste Valorization case is normalized by comparing the outcome of the green waste impact with the total environmental burden created in the Netherlands in 1997 due to economic activities. The result is called the normalized environmental effect scores for the valorization of green waste, and it is depicted in Table 5.7. The higher the score for a life impact category, the more negative the impact is on the environment. Negative environmental impact scores reduce the environmental burden by avoiding the environmental impact of an existing product (e.g., green waste compost can be used as a substitute for chemical fertilizer or dry green waste can be used for co-firing in biomass power plants).

In addition to the normalized environmental life cycle impact scores, some more managerial environmental performance indicators were calculated by SenterNovem (2008); they are depicted in Table 5.8.

Table 5.7 Normalized environmental effect scores ($*10^{-14}$) for Green Waste Valorization according SenterNovem (2008)

Life cycle impact category	Composting	Incineration with energy recuperation
Abiotic depletion	−13,455	−213,939
Global warming potential	−57,466	−171,041
Depletion ozone layer	−226	−1812
Photochemical oxidant formation	−4758	−12,253
Eco-toxicity (fresh water)	−4748	−9456
Eco-toxicity (terrestrial)	−6601	−32,430
Human toxicity	−723	−2218
Acidification	−18,386	−27,354
Eutrophication (aquatic)	−14,363	−4960
Eutrophication (terrestrial)	−16,460	−8460
Biodiversity	−7320	−27,577
Life support	−9200	−25,911

Table 5.8 Managerial environmental performance indicators (SenterNovem 2008)

	Composting	Incineration with energy recuperation
Land use [m^2 * year]	−2.11	−5.95
Final waste [kg]	−0.88	−37.4
Energy use [MJ]	−221	−5710
Water consumption (liter)	−33.4	−7280

Table 5.9 Weighted effect scores (* 10^{-12}) and significant differences (SenterNovem 2008)

	Composting	Incineration with energy recuperation
All environmental concerns equally weighted	−881	−3731
All life cycle impact categories equally weighted	−1537	−5374
Only greenhouse gasses taken into account	−575	−1710

4.4 Interpretation

Based on the sensitivity analysis that was carried out, variations in environmental impact can be compared across different scenarios. The results are listed in Table 5.9. Since lower scores reflect a better environmental outcome, the table values indicate that incineration with energy recuperation should be preferred in all scenarios because it has a better environmental outcome. That is due to the avoided emissions by both coal-fired power plants and heat produced by natural gas. The negative environmental impact of composting is due to the avoidance of chemical fertilizers and the transportation of these fertilizers.

5 LCA Used as an Input in MCDA

There is a growing consensus about the need to expand the ISO 14040+44 LCA framework by integrating and connecting it with other concepts and methods to increase its usefulness for sustainability decision making (Vadenbo et al. 2014; Banasik et al. 2016; Jacquemin et al. 2012; Jeswani et al. 2010). The review of Jacquemin et al. (2012) identifies that LCA results are often used as an input for the multi-objective optimization of processes. Practitioners use LCA to obtain a life cycle inventory then insert these results in an optimization model. It also shows that most of the LCA studies undertaken consider the unit processes as black boxes and build the inventory analysis using fixed operating conditions. According to Jeswani et al. (2010), multiple-criteria decision analysis (MCDA) could help broaden LCA to consider, among other things, social and economic impacts. This was done for the Green Waste Valorization case study, as explained in Sect. 6 of Chap. 1. Furthermore, an MCDA can be used to interpret the results of an LCA; they come in different units of measurements and often show goal conflicts (Jeswani et al. 2010).

Chapter 6
Quantifying the Social Impact

The last pillar of the triple bottom line to be quantified is the "people" pillar, which describes the social or societal impact. As discussed in Chap. 1, the triple bottom line framework requires that all three sustainability dimensions are quantified.

A commonly accepted definition for the social dimension is not yet available, mainly because there is no consensus on the meaning of the term 'social' (Lethonen 2004). The social dimension is immaterial and, therefore, difficult to analyze quantitatively (Lethonen 2004; Munda 2004). Since many social indicators cannot be quantified, qualitative ranking and scoring are currently used alongside quantitative measures (Klöpffer 2008). A popular multi-criteria decision making (MCDM) method that can be used to quantify such qualitative comparisons is the analytic hierarchy process (AHP) (Saaty 1980). The AHP method requires the social criteria of interest to be selected and rated using pairwise comparisons.

In this chapter, we introduce AHP as a methodology to quantify the social impact. First, we apply AHP on a simple, non-business process problem before applying it to the Green Waste Valorization study case introduced in Chap. 1.

1 Introduction to the AHP

AHP is a theory and methodology for relative measurement (Brunelli 2015). Relative measurement focuses on how things compare to each other, i.e., how proportions are related to each other. It suits measurements where it is not possible to have exact measurements of quantities and problems where the best alternative has to be chosen from a finite set of alternatives. The social impact of options is expressed in terms of how these options could be compared using criteria like safety, comfort, ease of use, perceived impact, etc.

D. Inghels, *Introduction to Modeling Sustainable Development in Business Processes*, https://doi.org/10.1007/978-3-030-58422-1_6

In general, AHP can be applied to decision making problems with one goal and a finite set of alternatives $X = \{x_1, \ldots, x_n\}$ from which decision makers are usually asked to select the best one (Brunelli 2015).

In this section, we introduce AHP using a simple, non-business related case study: a family consisting of four people (father, mother, and two children) wants to travel from Amsterdam to Rome, but has not decided yet which mode of transportation to use. Our family can choose between a car, train, or airplane. Their goal is to select the mode of transportation that gives them the highest overall satisfaction, which is determined by two criteria: cost price and comfort.

Mathematically, the set of alternatives is represented by $X = \{x_1, x_2, x_3\} = \{\text{car, train, airplane}\}$. The set of decision criteria is represented by $C = \{c_1, c_2\} = \{\text{price, comfort}\}$.

1.1 Pairwise Comparison of Criteria and Alternatives

The first step in AHP is to depict the problem using a decision tree with, in its simplest form, the goal of the decision problem on top and the alternatives to be selected below (see Fig. 6.1). Decision criteria will be added to the decision tree in the next step.

In an analytic hierarchy process, the decision maker has to formally assign a weighting factor, w_i, to the set of alternatives X. The sum of all the weighting factors must be equal to one. The set of weighting factors are combined in the weight vector W. Suppose that in the case of the selection of the preferred holiday transportation mode, with only three alternatives to consider, the weight vector W looks like this:

$$W = \begin{bmatrix} w_1 \\ w_2 \\ w_3 \end{bmatrix} = \begin{bmatrix} 0.08 \\ 0.19 \\ 0.73 \end{bmatrix} \tag{6.1}$$

The weight vector is often denoted in its transposed notation:

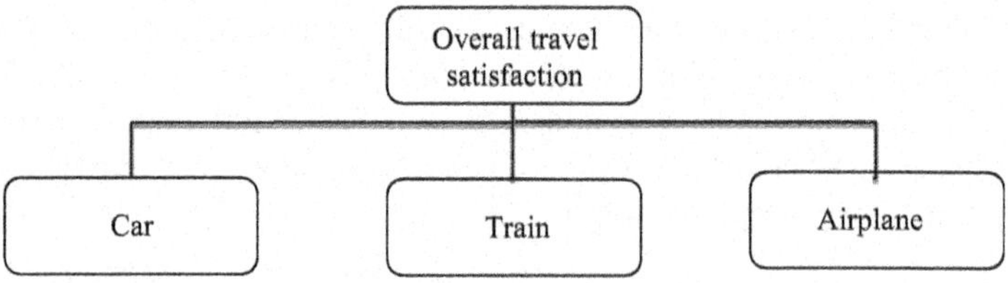

Fig. 6.1 Decision tree (goal and set of alternatives) for the selection of the holiday transportation mode

$$W = (w_1, w_2, w_3)^T = (0.08, 0.19, 0.73)^T$$

In general, the weight vector W for a set X with n alternatives is represented by $W = (w_1, w_2, \ldots, w_n)^T$.

Each value w_i in the weight vector corresponds with the relative preference assigned to the alternative x_i. In this case, the highest relative preference is assigned to the airplane (73%), followed by the train (19%), and car (8%). In other words, the preference ranking for the holiday transportation mode is airplane > train > car.

The decision making process in this simple case does not require a mathematical approach, but if the number of alternatives rises and decision criteria that must be applied to all the alternatives are taken into account, then making a decision becomes more complicated. For more complex problems, it is difficult to derive a preference ranking while taking all the set of alternatives and criteria into account. This is where the power of the AHP methodology comes in handy. AHP is based on the pairwise comparison of criteria and alternatives. This is much easier for humans to perform than merely assigning rating preferences taking all aspects into account at once.

A pairwise comparison allows the decision maker to concentrate each of his judgments on the comparison of two alternatives or criteria. This approach breaks a complex problem down by comparing a group of smaller subproblems or using pairwise comparisons. The pairwise comparisons are all gathered in the pairwise comparison matrix A, which is, in general, a n x n matrix for which n elements (alternatives, criteria) are compared.

In the holiday transportation mode case, the airplane is strongly preferred over the car, and the train is moderately preferred over the car (remember the preference is airplane > train > car). This is expressed in weighting preferences with w_1 being the weighting preference of the car option, w_2 of the train option, and w_3 of the airplane option. The pairwise comparison of these weighting preferences results in the relative preference of the alternatives. Let's use the following relative preferences: airplane versus car $w_3/w_1 = 7$, train versus car: $w_2/w_1 = 3$ and car versus car: $w_1/w_1 = 1$.

Additionally, the airplane option is strongly preferred over the train option, as expressed in the weighting factors $w_3/w_2 = 5$. It goes without saying that if the airplane is strongly preferred over the car ($w_3/w_1 = 7$), then the car is not strongly preferred over the airplane, as expressed with the weighting factors $w_1/w_3 = 1/7$.

The scores of the mutually compared alternatives are collected in a matrix, called the *comparison matrix A*.

According to Saaty (1980), all elements a_{ij} in comparison matrix A express the degree of preference of alternatives x_i to x_j. Therefore, each element of comparison in matrix A should approximate the ratio between the two weights:

$$a_{ij} \approx \frac{w_i}{w_j}; \forall i, j \tag{6.2}$$

Table 6.1 The fundamental AHP scale according Saaty and Vargas (1991)

Scale (i.e., intensity of importance)	Definition	Explanation
1	Equal importance	Two activities contribute equally to the objective
2	Weak	
3	Moderate importance	Experience and judgment slightly favor one activity over another
4	Moderate plus	
5	Strong importance	Experience and judgment strongly favor one activity over another
6	Strong plus	
7	Very strong or demonstrated importance	An activity is favored very strongly over another; its dominance demonstrated in practice
8	Very, very strong	
9	Extreme importance	The evidence favoring one activity over another is of the highest possible order of affirmation

For the holiday transportation mode case, the comparison matrix A is represented in (6.3). For convenience, we have indicated the positions of the alternatives x_1, x_2, and x_3 within A.

$$A = \begin{array}{c} \begin{matrix} x_1 & x_2 & x_3 \end{matrix} \\ \begin{bmatrix} \frac{w_1}{w_1} & \frac{w_1}{w_2} & \frac{w_1}{w_3} \\ \frac{w_2}{w_1} & \frac{w_2}{w_2} & \frac{w_2}{w_3} \\ \frac{w_3}{w_1} & \frac{w_3}{w_2} & \frac{w_3}{w_3} \end{bmatrix} \end{array} \approx \begin{bmatrix} \frac{1}{1} & \frac{1}{3} & \frac{1}{7} \\ \frac{3}{1} & \frac{1}{1} & \frac{1}{5} \\ \frac{7}{1} & \frac{5}{1} & \frac{1}{1} \end{bmatrix} = \begin{bmatrix} 1 & \frac{1}{3} & \frac{1}{7} \\ 3 & 1 & \frac{1}{5} \\ 7 & 5 & 1 \end{bmatrix} \begin{matrix} \rightarrow x_1 \\ \rightarrow x_2 \\ \rightarrow x_1 \end{matrix} \tag{6.3}$$

In the first row, the relative preferences for alternative x_1 compared to x_1, x_2, and x_3 are expressed using their weighting factors w_1, w_2, and w_3. In the second row, the relative preferences for alternative x_2 are compared to x_1, x_2, and x_3, and in the third row, this takes place for alternative x_3.

In order to make paired comparisons in a structured way, a scale has been developed by Saaty and Vargas (1991) that primarily uses odd numbers. The even numbers are used as intermediate values (see Table 6.1). Over time, other scales were developed based on additional insights (Brunelli 2015).

It can be helpful to present the pairwise comparison graphically on a scale, as depicted in Fig. 6.2. For each pairwise comparison, a box on the left side of the scale should be ticked if the left-side alternative is preferred over the right-side alternative. By contrast, if the right-side alternative is preferred above the left-side alternative, the box on the right side should be ticked.

This graphical representation can be used to transfer the pairwise comparison data into comparison matrix A. For example, take the first row of matrix A from (6.3),

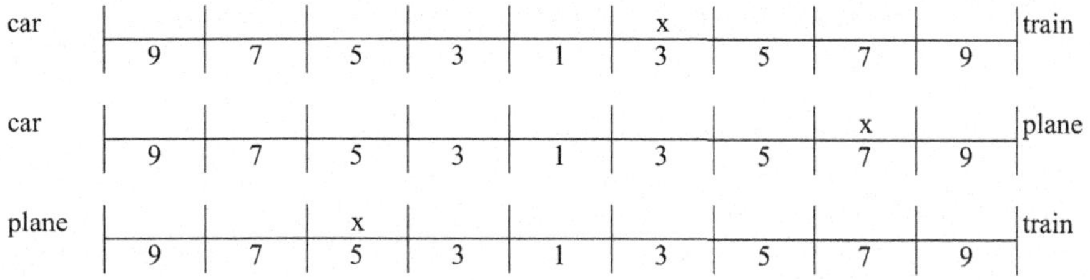

Fig. 6.2 Graphical representation of the pairwise comparison between preferences of the holiday mode of transport example

Table 6.2 Number of comparisons in function of the number of alternatives

Number of alternatives n	1	2	3	4	5	6	7	n
Number of comparisons	0	1	3	6	10	15	21	$\frac{n \cdot (n-1)}{2}$

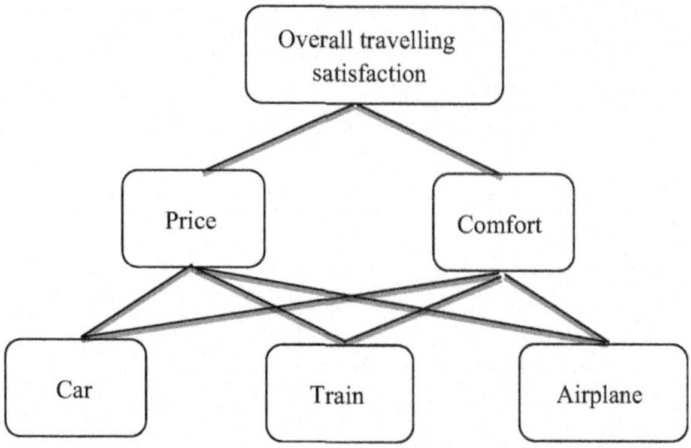

Fig. 6.3 Extended decision tree (goal, criteria and alternatives) for the selection of the holiday transportation mode

which represents the pairwise comparison of the option 'car' with the options 'train' and 'airplane'. Numbers ticked on the right side of the scale have to be entered with their reciprocal value in comparison matrix A. Numbers ticked on the left side can be entered as such.

The number of comparisons to be made in a comparison matrix depends upon its size n. Table 6.2 depicts how many comparisons should be made for a $n \times n$ comparison matrix A:

In most decision problems, a set of criteria $C = \{c_1, c_2, \ldots, c_n\}$ is added and applied to the different alternatives. In the case of the selection of the preferred holiday transportation mode, the selection criteria are $C = \{c_1, c_2\} = \{\text{price, comfort}\}$.

The extended decision tree encompassing the goal, decision criteria, and alternatives is depicted in Fig. 6.3.

Notc that for each of the criteria, in this case price and comfort, the related relative preferences may differ for the alternatives. Therefore, the next step in the analytical hierarchy process is to build a comparison matrix for each criterion. These comparison matrices are denoted by A_i, whereby the index i in subscript refers to the specific criterion c_i for which the alternatives are compared pairwise. The same goes for the related priority vector $W_{i.}$

In the holiday transportation mode case, a comparison matrix for the criteria Price, A_p, and Comfort, A_c, has to be constructed. Both contain the pairwise comparison of the three alternatives according to each criterion.

Suppose that the outcome is the following for the pairwise comparison matrix A_p and the related priority vector W_p:

$$A_p = \begin{bmatrix} 1 & 3 & 5 \\ \frac{1}{3} & 1 & 3 \\ \frac{1}{5} & \frac{1}{3} & 1 \end{bmatrix}; W_p = \begin{bmatrix} 0.64 \\ 0.26 \\ 0.10 \end{bmatrix} \tag{6.4}$$

The priority vector W_p shows the preference ranking of the mode of transportation alternatives with respect to 'price'. Since transportation by car is the cheapest mode of transportation for a family of four, it is the preferred option, followed by the options train and airplane.

The pairwise comparison for the criterion 'Comfort' is expressed in (6.5). Since traveling by airplane is the most convenient for a family of four, it gets the highest score in the corresponding weight vector W_c:

$$A_C = \begin{bmatrix} 1 & \frac{1}{3} & \frac{1}{5} \\ 3 & 1 & \frac{1}{3} \\ 5 & 3 & 1 \end{bmatrix}; W_C = \begin{bmatrix} 0.10 \\ 0.26 \\ 0.64 \end{bmatrix} \tag{6.5}$$

To make a sound comparison, the relative importance of the criteria should also be pairwise compared. The pairwise comparison matrix of the criteria is expressed in the comparison matrix $\widehat{A}$. The associated priority vector for the criteria is denoted as $\widehat{W}$. Price is considered twice as important as comfort in the holiday mode of transportation case. Therefore, the pairwise comparison matrix for the criteria Price and Comfort is expressed in (6.6) as:

$$\widehat{A} = \begin{bmatrix} 1 & 2 \\ \frac{1}{2} & 1 \end{bmatrix}; \widehat{W} = \begin{bmatrix} 0.67 \\ 0.33 \end{bmatrix} \begin{matrix} P \rightarrow \widehat{w}_1 \\ C \rightarrow \widehat{w}_2 \end{matrix} \tag{6.6}$$

The final preference ranking related to the three alternative holiday modes of transportation 'car', 'train' and 'airplane' according to the two-decision criteria 'price' and 'comfort' are represented by the following weighted arithmetic mean:

$$W = \widehat{w}_1 \cdot W_P + \widehat{w}_2 \cdot W_C = 0.67 \cdot \begin{bmatrix} 0.64 \\ 0.26 \\ 0.10 \end{bmatrix} + 0.33 \cdot \begin{bmatrix} 0.10 \\ 0.26 \\ 0.64 \end{bmatrix} = \begin{bmatrix} 0.46 \\ 0.26 \\ 0.28 \end{bmatrix} \tag{6.7}$$

The final conclusion, taking the options and alternatives of the holiday mode of transportation into account, is that the preferred mode of transportation to travel from Amsterdam to Rome is by car (46%) followed by airplane (28%) and train (26%). Note that this differs from the earlier weight vector expressed in (6.1), where only the alternatives were considered.

1.2 Generalization

In summary, the analytical hierarchy process to select the preferred alternative x_i, from a set of alternatives X, taking a set of criteria C into account, consists of four consecutive stages:

1. Define the hierarchy goal, decision criteria $C = \{c_1, c_2, \ldots, c_n\}$ and alternatives $X = \{x_1, x_2, \ldots, x_m\}$.
2. Compute the pairwise comparison matrix $\widehat{A}$ of the set of decision criteria C, and the associated priority vector $\widehat{W}$.
3. Compute the comparison matrices of the alternatives in set X, A_i, for each criterion i of set C and the associated priority vectors W_i.
4. Rank the options and conclude based on the preferences of the alternatives resulting in priority vector W of the alternatives taking all criteria and their relative importance into account.

For any pairwise comparison matrix A consisting of elements a_{ij} in row i and column j, the pairwise comparison ratio of each variable in row i compared to the variable in column j has to be entered using the scale depicted in Table 6.1.

Comparison matrix A is a $n \times n$ matrix with n the number of things (i.e. criteria or alternatives) to be compared as expressed in (6.8):

$$A = \begin{bmatrix} a_{11} & \cdots & a_{1n} \\ \vdots & \ddots & \vdots \\ a_{n1} & \cdots & a_{nn} \end{bmatrix} \tag{6.8}$$

For the comparison matrix A, the following rules apply:

- The diagonal values $a_{11}, a_{22}, \ldots, a_{nn}$ of matrix A are always 1.
- Only the upper triangular matrix values $a_{11}, a_{12}, \ldots a_{1n}, a_{22}, a_{23}, \ldots a_{2n}$, etc. should be filled in as depicted in (6.9):

$$A = \begin{bmatrix} 1 & a_{12} & \dots & a_{1n} \\ & 1 & \dots & a_{2n} \\ & & 1 & \dots \\ & & & 1 \end{bmatrix} \tag{6.9}$$

- The following rules apply to the values in the upper triangular matrix ($a_{12}, \ldots, a_{1n}, a_{23} \ldots a_{2n}, \ldots$):
  - If the judgment is ticked on the left side of 1 on the graphical comparison scale (see Fig. 6.2), place the actual judgment value in matrix A.
  - If the judgment is ticked on the right side of 1 on the graphical comparison scale, place the reciprocal value in matrix A
- The lower triangular matrix gets the reciprocal values of the upper triangular matrix assigned using the following expression (6.10)

$$a_{ji} = \frac{1}{a_{ij}} \tag{6.10}$$

Comparison matrix A depicted in (6.8) can be completed as follows:

$$A = \begin{bmatrix} 1 & a_{12} & \dots & a_{1n} \\ \frac{1}{a_{12}} & 1 & \dots & a_{2n} \\ \vdots & \dots & 1 & \dots \\ \frac{1}{a_{1n}} & \frac{1}{a_{2n}} & \dots & 1 \end{bmatrix} \tag{6.11}$$

For the sake of clarity, we list the notations used in this chapter on AHP:

X: set of alternatives.

C: set of decision criteria.

A: pairwise comparison matrix of all alternatives X.

A_i: pairwise comparison matrix of all alternatives X with respect to criterion c_i.

$\widehat{A}$: pairwise comparison matrix of the decision criteria C.

$\widetilde{A}$: normalized approximated pairwise comparison matrix of the decision criteria C.

W: priority vector of alternatives X taking the relative weight of all criteria into account; normalized principal eigenvector of matrix A.

W_i: priority vector of the pairwise comparison matrix for all alternatives X with respect to criterion c_i; normalized principal eigenvector of matrix A_i.

$\widehat{W}$: priority vector of the pairwise comparison matrix of the decision criteria C; normalized principal eigenvector of matrix $\widehat{A}$.

$\widetilde{W}$: approximated priority vector of the pairwise comparison matrix of the decision criteria C; normalized principal eigenvector of matrix $\widetilde{A}$

$\overline{W}$: principal eigenvector

In the next section, we will discuss how to calculate the priority vector associated with a pairwise comparison matrix.

2 Computing the Priority Vector

In this section, we briefly discuss two exact methods for deriving the priority vector W related to a comparison matrix A: the *eigenvector* method proposed by Saaty (1980), the founding father of AHP, and the *geometric mean* method proposed by Crawford and Williams (1985).

2.1 The Eigenvector Method

2.1.1 Computation of the Priority Vector W

The priority vector W of a comparison matrix A is defined as being the *normalized principal eigenvector* of matrix A.

We illustrate the computation of priority vector W with the eigenvector method using the comparison matrix (6.3) as an example. The first step is to calculate the eigenvalues for comparison matrix A. For the squared n x n comparison matrix A, each eigenvalue is a scalar λ such that

$$|A - \lambda \cdot I| = 0 \Leftrightarrow \begin{vmatrix} 1-\lambda & \frac{1}{3} & \frac{1}{7} \\ 3 & 1-\lambda & \frac{1}{5} \\ 7 & 5 & 1-\lambda \end{vmatrix} = 0 \tag{6.12}$$

where I is the $n \times n$ identity matrix defined by $a_{ii} = 1$ and $a_{ij} = 0 \; \forall i, j \in A$.

The *principal eigenvector* $\overline{W}$, is the eigenvector corresponding to the highest eigenvalue λ_{max} of comparison matrix A and is determined by solving (6.13):

$$A \cdot \overline{W} = \lambda_{max} \cdot \overline{W} \tag{6.13}$$

Solving (6.13) yields a characteristic polynomial equation of the n-order for the variable λ. In total, n eigenvalues are the roots of this equation. They can be real or complex values.

Applied to matrix (6.3), the characteristic polynomial is:

$$\lambda^3 - 3 \cdot \lambda^2 - 0.61 = 0 \tag{6.14}$$

It has one real eigenvalue $\lambda = 3.065$ and two complex eigenvalues $\lambda = \{-0.03244 + 0.44477 \cdot \hat{\imath}, -0.03244 - 0.44477 \cdot \hat{\imath}\}$. Note that both eigenvalues and eigenvectors can be derived using specialized software.

Since there is only one real eigenvalue for (6.14), this eigenvalue is also the principal eigenvalue λ_{max}.

The eigenvector corresponding with the principal eigenvalue $\lambda_{max} = 3.065$ is:

$$\overline{W} = (0.107, 0.248, 0.963)^{\mathrm{T}} \tag{6.15}$$

The sum of all elements of the principal eigenvector is $0.107 + 0.248 + 0.963 = 1.318$

Priority vector W is obtained by dividing the elements of the principal eigenvector $\overline{W}$ by the sum of its elements. Therefore, we divide each element in of the principal eigenvector $\overline{W}$ expressed in (6.15) by 1.318. The outcome yields priority vector W of matrix A:

$$W = \begin{bmatrix} \frac{0.107}{1.318} \\ \frac{0.248}{1.318} \\ \frac{0.963}{1.318} \end{bmatrix} = \begin{bmatrix} 0.081 \\ 0.188 \\ 0.730 \end{bmatrix} \tag{6.16}$$

Note that the outcome of (6.16) is identical to (6.1). The other priority vectors Wi and $\widehat{W}$ can be computed the same way.

2.1.2 Computation of the Consistency Index *CI*

Since this pairwise comparison of priorities by decision makers is generally not consistent, by definition, the AHP method allows for some inconsistency. The level of (in)consistency is checked by calculating the *consistency index*. The consistency index is computed from the principal eigenvalue λ_{max} of comparison matrix A. Saaty (2006) defines the consistency index, CI, as expressed in (6.17):

$$CI = \frac{\lambda_{max} - n}{n - 1} \tag{6.17}$$

with n denoting the rank of the n x n comparison matrix and λ_{max} denoting the principal eigenvalue of A.

Applied to comparison matrix A of (6.3), $\lambda_{max} = 3.065$ and $n = 3$. Consequently,

Table 6.3 Radom consistency index *RI* in function of the number of comparisons *n*

n	1	2	3	4	5	6	7	8	9	10
RI	0	0	0.58	0.89	1.12	1.24	1.32	1.41	1.45	1.49

$$CI = \frac{3.065 - 3}{3 - 1} = 0.033$$

The consistency index *CI* has to be compared to a value derived by generating random reciprocal matrices *RI* of the same size (see Table 6.3) to compute a consistency ratio *CR* (6.18). The random consistency index, *RI*, depends on the size of the criteria comparison matrix.

If the consistency ratio $CR < 0.1$, the comparison is assumed to be consistent (Saaty 2006).

$$CR = \frac{CI}{RI} \tag{6.18}$$

Applied to a 3x3 matrix *A*, the $RI = 0.58$. Therefore, $CR = \frac{0.033}{0.58} = 0.057 < 0.1$ so the element values of comparison matrix *A* are assumed to be consistent.

2.2 The Geometric Mean Method

Another exact method for deriving the priority vector *W* is the geometric mean method proposed by Crawford and Williams (1985). Using this method, each element w_i of priority vector *W* is determined as the geometric mean of the elements of the respective row divided by a normalization term as expressed in (6.19).

$$w_i = \frac{\left(\prod_{j=1}^{n} a_{ij}\right)^{1/n}}{\sum_{i=1}^{n}\left(\prod_{j=1}^{n} a_{ij}\right)^{1/n}} \tag{6.19}$$

where:

$n =$ is rank of the n x n matrix *A*

$$\prod_{j=1}^{n} a_{ij} = a_{i1} \cdot a_{i2} \cdots a_{ij}$$

We apply this method to the same comparison matrix $A = \begin{bmatrix} 1 & \frac{1}{3} & \frac{1}{7} \\ 3 & 1 & \frac{1}{5} \\ 7 & 5 & 1 \end{bmatrix}$ as in (6.3).

The first weighting factor w_1 of the priority vector associated with comparison matrix A is

$$w_1 = \frac{\left(1 \cdot \frac{1}{3} \cdot \frac{1}{7}\right)^{1/3}}{\left\{\left(1 \cdot \frac{1}{3} \cdot \frac{1}{7}\right)^{1/3} + \left(3 \cdot 1 \cdot \frac{1}{5}\right)^{1/3} + (7 \cdot 5 \cdot 1)^{1/3}\right\}} = 0.081 \tag{6.20}$$

Weighting factors w_2 and w_3 are derived the same way:

$$w_2 = \frac{\left(3 \cdot 1 \cdot \frac{1}{5}\right)^{1/3}}{\left\{\left(1 \cdot \frac{1}{3} \cdot \frac{1}{7}\right)^{1/3} + \left(3 \cdot 1 \cdot \frac{1}{5}\right)^{1/3} + (7 \cdot 5 \cdot 1)^{1/3}\right\}} = 0.188 \tag{6.21}$$

$$w_3 = \frac{(7 \cdot 5 \cdot 1)^{1/3}}{\left\{\left(1 \cdot \frac{1}{3} \cdot \frac{1}{7}\right)^{1/3} + \left(3 \cdot 1 \cdot \frac{1}{5}\right)^{1/3} + (7 \cdot 5 \cdot 1)^{1/3}\right\}} = 0.730 \tag{6.22}$$

Note that the sum of the weighting factors $\sum_{i=1}^{n} w_i$ equals 1 taking rounding into account:

$$\sum_{i=1}^{3} w_i = 0.081 + 0.188 + 0.730 = 1 \tag{6.23}$$

The result is precisely the same priority vector as expressed in (6.16).

The other priority vectors can be computed for all the other comparison matrices using the same method. Now that we know how to apply the AHP on a simple case, we are ready to apply it to a more complex case: determining the social impact of the Green Waste Valorization case as described by Inghels et al. (2016b).

3 Determining the Social Impact of the Green Waste Valorization Case

Figure 6.4 represents the various options for the recovery of green waste embedded in the aerobic composting process of green waste. It represents a generic mass balance of the incoming green waste for an instance in which all green waste is

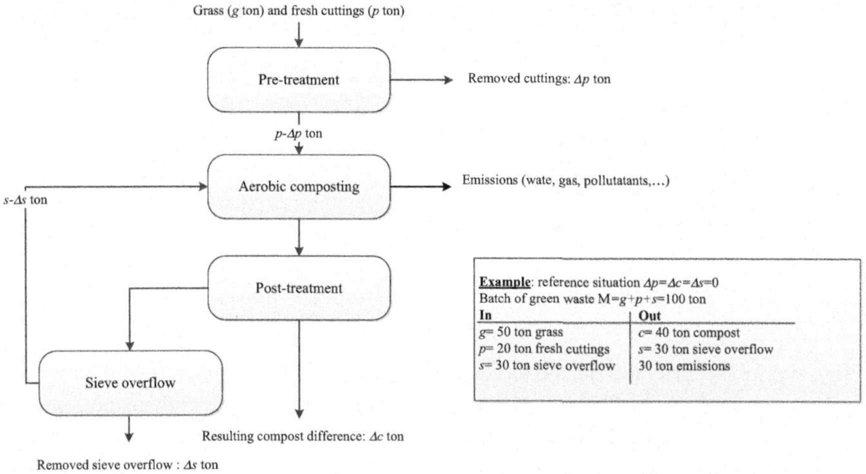

Fig. 6.4 Overview of the aerobic composting process with possible separation of fresh wooden cuttings and sieve overflow for energy valorization according Inghels et al. (2016b)

composted. The typical mass input of a batch of green waste is composed of 50% grass, 30% sieve overflow (a remainder of the previous composting run), and 20% fresh wooden cuttings. This leads to a typical mass output of 40% compost, 30% sieve overflow, and 30% emissions such as water, gasses, and pollutants.

In this section, we want to assess whether investements in separating sieve overflow and/or fresh wooden cuttings from a batch of green waste can be considered as more sustainable in terms of societal impact than composting green waste exclusively. This assessment is based on Inghels et al. (2016b)

Let:

$p =$ amount of fresh wood cuttings (from pruning) in a batch of green waste [ton]

$s =$ amount of sieve overflow in a batch of green waste [ton]

$g =$ amount of grass in a batch of green waste [ton]

$c =$ resulting amount of compost [ton]

The four Green Waste Valorization alternatives denoted $\alpha_1 \ldots \alpha_4$ are:

α_1: exclusive composting, which is the reference scenario

α_2: additional to α_1 removing of fresh wooden cuttings after composting by sieving the compost, which is more labor intensive

α_3: additional to α_2 removing a part of the fresh wooden cuttings before composting to be sold as biomass for power plants

α_4: additional to α_1 removing a part of the fresh wooden cuttings before composting

The four alternatives $k = \{1, 2, 3, 4\}$ are summarized in Table 6.4:

For the quantification of the relative social impact associated with the four alternatives listed in Table 6.4, we use AHP. Therefore, the relevant social impact criteria need to be defined first. Secondly, relative priority levels for the $l = 4$ selected social impact criteria (safety (SA), local employment (LE), job enrichment

Table 6.4 Overview of the four investment alternatives to be considered for the valorization of green waste adapted from Inghels et al. (2016b)

Alternative k	Description	Sieve overflow separated after composting	Fresh cuttings separated during pre-treatment
α_1	Exclusive composting (reference scenario)	☒	☒
α_2	Only sieve overflow	☑	☒
α_3	Pre-treatment & sieve overflow	☑	☑
α_4	Only pre-treatment	☒	☑

(JE), and job security (JS)) need to be determined by assigning normalized weighting factors ω_l to each. Thirdly, for each criterion, the normalized weight factors ω_{lk} for social impact criterion l on alternative k are derived. The relative weight of each criterion l per alternative k is used to determine the social impact for each alternative k. The higher the impact score, the more social importance is linked to that alternative.

Let:

$\alpha_k =$ alternative k ($\alpha_k = 1$ if alternative k is selected, otherwise $\alpha_k = 0$)

$\omega_l =$ normalized weight factor for social impact criterion l

$\omega_{lk} =$ weight factor for social impact criterion l on alternative k

The goal is to formulate a societal objective function Z_5 (see section 6.6 in Chap. 1) that can be used in a SOOP or a MOOP as expressed in (6.24). Additional constraints are that the green waste recovery alternatives are mutually exclusive (6.25) and that the variable used for expressing the alternatives k, α_k, is binary (6.26).

$$\text{Maximize } Z_5 = \sum_{l=1}^{l} \sum_{k=1}^{k} \varpi_l \cdot \varpi_{lk} \cdot \alpha_k \tag{6.24}$$

s.t.

$$\sum_{k=1}^{k} \alpha_k \tag{6.25}$$

$$\alpha_k \in \{0, 1\} \tag{6.26}$$

The social impact objective function (6.24) is the final outcome of the AHP assessment; it assigns a social impact factor to the $k = 4$ alternatives under investigation. The alternative k with the highest social impact factor is the most sustainable from the social impact objective point of view. The decision tree, encompassing the goal or focus, decision criteria, and alternatives are depicted in Fig. 6.5.

The next step in the analytic hierarchy process is the computation of priority vectors. To this end, we must first construct the comparison matrices for the pairwise comparison of the four alternatives for each of the four criteria, A_{SA}, A_{LE}, A_{JE} and

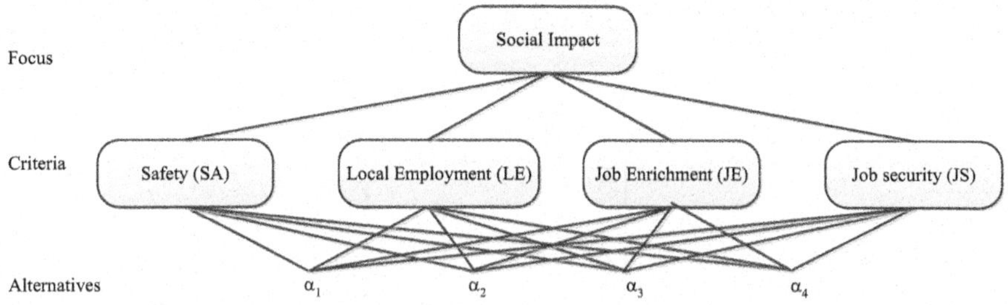

Fig. 6.5 Decision tree to assess the social impact of the four investment alternatives for the Green Waste Valorization case. (Source: Inghels et al., 2016b)

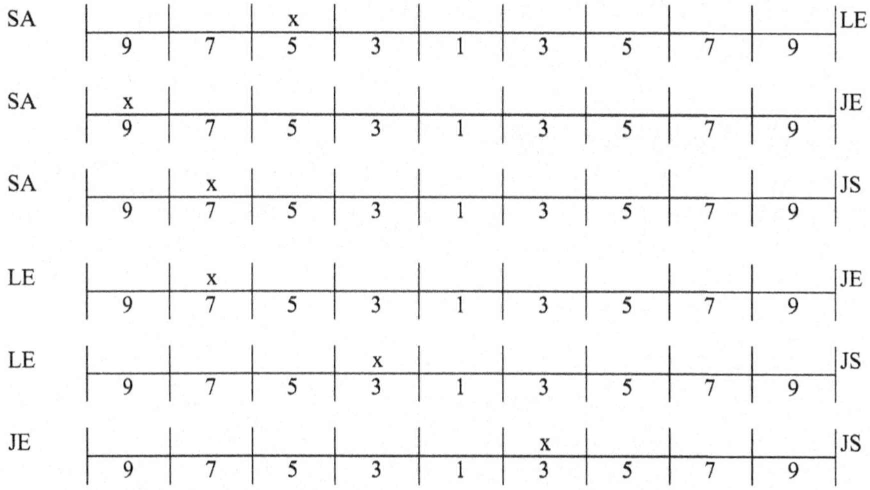

Fig. 6.6 Pairwise ranking of the decision criteria of the Green Waste Valorization case

A_{JS}, and the comparison matrix for the pairwise comparison of the decision criteria A.

The decision criteria are pairwise ranked by relative importance on a scale ranging from 1 to 9 using the comparison scale of Table 6.1. For example, safety (SA) is found to be more important than local employment (LE). Therefore, the box on the left side of the scale is ticked. By contrast, job security (JS) is found to be more important than job enrichment (JE), which results in the box on the right side of the scale being ticked. For the Green Waste Valorization case, the relative importance comparison of the social impact criteria have been scored as follows: $w_{SA}/w_{LE} = 5/1$, $w_{SA}/w_{JE} = 9/1$, $w_{SA}/w_{JS} = 7/1$, $w_{LE}/w_{JE} = 7/1$, $w_{LE}/w_{JS} = 3/1$, $w_{JE}/w_{JS} = 1/3$. The pairwise comparison of the decision criteria is depicted in Fig. 6.6.

The pairwise ranking scores of the decision criteria are transferred into comparison matrix A in (6.27). In the first row of the upper triangular matrix, the SA ranking scores versus SA, LE, JE, and JS are filled in. In the second row, the LE ranking scores versus LE, JE, and JS are filled in, etcetera.

$$
\begin{array}{c}
\text{Col}\,j\ \ \text{SA LE JE JS} \\
\downarrow
\end{array}
\qquad \text{Row } i
$$

$$A = \begin{bmatrix} 1 & 5 & 9 & 7 \\ \frac{1}{5} & 1 & 7 & 3 \\ \frac{1}{9} & \frac{1}{7} & 1 & \frac{1}{3} \\ \frac{1}{7} & \frac{1}{3} & 3 & 1 \end{bmatrix} \begin{matrix} \leftarrow \text{SA} \\ \text{LE} \\ \text{LE} \\ \text{JS} \end{matrix} \tag{6.27}$$

Next, we will determine the priority vector W of comparison matrix A using the eigenvector method and the geometric mean method.

3.1 The Eigenvector Method

For the eigenvector method, the eigenvalues λ of comparison matrix A are determined by solving the determinant $|A - \lambda \cdot I| = 0$. I is the $n\ x\ n$ identity matrix (in this case, a 4x4 matrix). I is expressed in (6.28).

$$I = \begin{bmatrix} 1 & 0 & 0 & 0 \\ 0 & 1 & 0 & 0 \\ 0 & 0 & 1 & 0 \\ 0 & 0 & 0 & 1 \end{bmatrix} \tag{6.28}$$

$$|A - \lambda \cdot I| = \begin{vmatrix} 1-\lambda & 5 & 9 & 7 \\ \frac{1}{5} & 1-\lambda & 7 & 3 \\ \frac{1}{9} & \frac{1}{7} & 1-\lambda & \frac{1}{3} \\ \frac{1}{7} & \frac{1}{3} & 3 & 1-\lambda \end{vmatrix} = 0 \tag{6.29}$$

Solving the expression (6.29) yields the characteristic polynomial expression (6.30):

$$\lambda^4 - 4 \cdot \lambda^3 - 3.581 \cdot \lambda - 0.252 = 0 \tag{6.30}$$

This results in a set of four eigenvalues consisting of two real eigenvalues and two complex eigenvalues: real eigenvalues: {−0.070, 4.206}; complex eigenvalues: {−0.068 + 0.923·î, −0.068−0.923·î}.

The principal eigenvalue is $\lambda_{max} = 4.206$.

The principal eigenvector belonging to the principal eigenvalue $\lambda_{max} = 4.206$ is

$$\overline{W} = \begin{bmatrix} 0.939 \\ 0.312 \\ 0.060 \\ 0.130 \end{bmatrix} \tag{6.31}$$

The sum of all the elements of matrix $\overline{W}$ is $0.939 + 0.312 + 0.060 + 0.130 = 1.441$

The normalized principal eigenvector or priority vector W is determined by dividing all elements of (6.31) by the sum of the element 1.441. This results in priority vector W:

$$W = \begin{bmatrix} 0.652 \\ 0.217 \\ 0.042 \\ 0.090 \end{bmatrix} \tag{6.32}$$

The preferred ranking for the pairwise comparison of the decision criteria results in SA: 65%, LE: 22%, JE: 0.04%, and JS: 0.09%.

Next, we check the consistency of comparison matrix A:

$$\text{Since } \lambda_{max} = 4.206, \text{and } n = 4, \text{than } CI = \frac{4.206 - 4}{4 - 1} = 0.069. \tag{6.33}$$

In the event of a 4×4 matrix A, $RI = 0.89$ (See Table 6.3). Therefore $CR = \frac{CI}{RI} = \frac{0.069}{0.89} = 0.078 < 0.1$ which is assumed to be consistent.

In addition to the pairwise comparison of the decision criteria, the four alternatives α_k have to be pairwise compared for each of the four criteria safety (SA), local employment (LE), job enrichment (JE), and job security (JS).

We start with the pairwise comparison of the four alternatives for the decision criterion safety (SA).

For safety, the preference ranking of the four alternatives is $\alpha_1 > \alpha_4 > \alpha_3 > \alpha_2$. The pairwise comparison matrix A_{SA}, as agreed by green waste practitioners, is depicted in (6.34).

$$A_{SA} = \begin{bmatrix} 1 & 9 & 7 & 3 \\ \frac{1}{9} & 1 & \frac{1}{3} & \frac{1}{5} \\ \frac{1}{7} & 3 & 1 & \frac{1}{5} \\ \frac{1}{3} & 5 & 5 & 1 \end{bmatrix} \tag{6.34}$$

The largest eigenvalue for the pairwise comparison matrix A_{SA} is $\lambda_{max} = 4.176$. The corresponding principal eigenvector $\overline{w}_{SA}$ is:

$$\overline{w}_{SA} = \begin{bmatrix} 0.895 \\ 0.072 \\ 0.134 \\ 0.419 \end{bmatrix} \quad (6.35)$$

The sum of all the elements of this matrix is $0.895 + 0.072 + 0.134 + 0.419 = 1.52$. This allows us to calculate the priority vector W_{SA}, which shows the preference ranking of the pairwise comparison of the four alternatives on safety quantitatively.

$$W_{SA} = \begin{bmatrix} 0.589 \\ 0.047 \\ 0.088 \\ 0.276 \end{bmatrix} \quad (6.36)$$

Since $\lambda_{max} = 4.176$ and $n = 4$ then according to (6.17): $CI = \frac{4.176-1}{4-1} = 0.059$.

Therefore, following (6.18) and Table 6.3, $CR = \frac{CI}{RI} = \frac{0.059}{0,89} = 0.066 < 0.1$, which is assumed to be consistent.

The same process of pairwise comparison must take place for all the remaining criteria for each of the four alternatives.

Let's assume that the preferences stated by the green waste practitioners are as follows for LE, JE and JS:

LE: $\alpha_3 = \alpha_4 > \alpha_1 = \alpha_2$

JE: $\alpha_3 > \alpha_4 > \alpha_1 > \alpha_2$

JS: $\alpha_3 > \alpha_4 > \alpha_1 > \alpha_2$ (i.e., identical to JE)

The corresponding comparison matrices, principal eigenvalues, priority vectors, *CI,* and *CR* are:

$$A_{LE} = \begin{bmatrix} 1 & 1 & \frac{1}{7} & \frac{1}{7} \\ 1 & 1 & \frac{1}{7} & \frac{1}{7} \\ 7 & 7 & 1 & 1 \\ 7 & 7 & 1 & 1 \end{bmatrix}; \overline{w}_{LE} = \begin{bmatrix} 1 \\ 1 \\ 7 \\ 7 \end{bmatrix}; W_{LE} = \begin{bmatrix} 0.063 \\ 0.063 \\ 0.438 \\ 0.438 \end{bmatrix}; \lambda_{max} = 4, CI = 0;$$

$$CR = 0 < 0.1, \text{which is assumed to be consistent.} \quad (6.37)$$

$$A_{JE} = \begin{bmatrix} 1 & 3 & \frac{1}{7} & \frac{1}{5} \\ \frac{1}{3} & 1 & \frac{1}{9} & \frac{1}{5} \\ 7 & 9 & 1 & 3 \\ 5 & 5 & \frac{1}{3} & 1 \end{bmatrix}; \overline{w}_{JE} = \begin{bmatrix} 0.134 \\ 0.072 \\ 0.895 \\ 0.419 \end{bmatrix}; W_{JE} = \begin{bmatrix} 0.088 \\ 0.047 \\ 0.589 \\ 0.276 \end{bmatrix}; \lambda_{max} = 4.176,$$

$CI = 0.059;$

$$CR = 0.06 < 0.1, \text{which is assumed to be consistent.} \tag{6.38}$$

Similar to the pairwise comparison of the four alternatives for JS:

$$A_{JS} = \begin{bmatrix} 1 & 3 & \frac{1}{7} & \frac{1}{5} \\ \frac{1}{3} & 1 & \frac{1}{9} & \frac{1}{5} \\ 7 & 9 & 1 & 3 \\ 5 & 5 & \frac{1}{3} & 1 \end{bmatrix}; \overline{w}_{JS} = \begin{bmatrix} 0.134 \\ 0.072 \\ 0.895 \\ 0.419 \end{bmatrix}; W_{JS} = \begin{bmatrix} 0.088 \\ 0.047 \\ 0.589 \\ 0.276 \end{bmatrix}; \lambda_{max} = 4.176,$$

$CI = 0.059;$

$$CR = 0.06 < 0.1, \text{which is assumed to be consistent.} \tag{6.39}$$

Table 6.5 summarizes the priority vector outcome for the pairwise comparison of the four investement alternatives for each criterion. In the above row, the normalized weighting factors ω_l are listed between brackets for each criterion, which corresponds to the priority vector of the pairwise comparison matrix of the decision criteria. For example, safety (SA) has a normalized weighting factor of 0.65 derived from the corresponding priority vector (6.32). Per criterion, the normalized weight factors ω_{lk} for social impact criterion l on alternative k are listed in each column. The depicted values are the corresponding elements of the priority vectors (6.36) for criterion SA, (6.37) for LE, (6.38) for JE, and (6.39) for JS.

Table 6.5 Summary of the elements of the priority vectors on social impact for the Green Waste Valorization case investement alternatives

Normalized weight factors ω_{lk}	Criteria l (ω_l listed between brackets for each criterion)			
Alternative k	SA (0.65)	LE (0.22)	JE (0.04)	JS (0.09)
α_1	0.59	0.06	0.09	0.09
α_2	0.05	0.06	0.05	0.05
α_3	0.09	0.44	0.59	0.59
α_4	0.28	0.44	0.28	0.28

Table 6.6 Social impact preferences and preference ranking for the Green Waste Management case

Alternative k	Social impact	Preference ranking
α_1	0.40	1
α_2	0.05	4
α_3	0.23	3
α_4	0.26	2

Finally, the contribution of each alternative to the overall goal Z_5 regarding the social impact of the four investment alternatives, k, has to be calculated. For example, the social impact related to alternative α_2 is calculated as:

$$Z_{5,\,\alpha 2} = (0.65)\cdot(0.05) + (0.22)\cdot(0.06) + (0.04)\cdot(0.05) + (0.09)\cdot(0.05) \\ = 0.05$$

The social impact of all four investment alternatives, taking the pairwise preferences for the four decision criteria and alternatives into account, is calculated and summarized in Table 6.6.

The final AHP assessment outcome is that the first alternative α_1, i.e. exclusive composting, is preferred above (in declining order) α_4, α_3, and α_2.

If these results are used in a SOOP or MOOP, Z_5 in (6.24) can now be expressed as:

$$Z_5 = 0.40\cdot\alpha_1 + 0.05\cdot\alpha_2 + 0.23\cdot\alpha_3 + 0.26\cdot\alpha_4 \tag{6.40}$$

3.2 *The Geometric Mean Method*

The priority vectors derived in Sect. 6.3.1. can also be determined using the geometric mean method. We demonstrate that this results in the same outcome as the eigenvector method by applying the geometric method on comparison matrix A of the pairwise decision criteria represented in matrix A expressed in (6.27).

The elements of priority vector W, consisting of the weighting factor w_i, are computed using (6.19). The first weighting factor w_1 of the priority vector associated with comparison matrix A is:

$$w_1 = \frac{(1\cdot 5\cdot 9\cdot 7)^{1/4}}{\left\{(1\cdot 5\cdot 9\cdot 7)^{1/4} + \left(\frac{1}{5}\cdot 1\cdot 7\cdot 3\right)^{1/4} + \left(\frac{1}{9}\cdot\frac{1}{7}\cdot 1\cdot\frac{1}{3}\right)^{1/4} + \left(\frac{1}{7}\cdot\frac{1}{3}\cdot 3\cdot 1\right)^{1/4}\right\}} \\ = 0.65 \tag{6.41}$$

Weighting factors w_2, w_3, and w_4 can be computed the same way:

$$w_2 = \frac{\left(\frac{1}{5}\cdot 1\cdot 7\cdot 3\right)^{1/4}}{\left\{(1\cdot 5\cdot 9\cdot 7)^{1/4} + \left(\frac{1}{5}\cdot 1\cdot 7\cdot 3\right)^{1/4} + \left(\frac{1}{9}\cdot\frac{1}{7}\cdot 1\cdot\frac{1}{3}\right)^{1/4} + \left(\frac{1}{7}\cdot\frac{1}{3}\cdot 3\cdot 1\right)^{1/4}\right\}} = 0.22 \quad (6.42)$$

$$w_3 = \frac{\left(\frac{1}{9}\cdot\frac{1}{7}\cdot 1\cdot\frac{1}{3}\right)^{1/4}}{\left\{(1\cdot 5\cdot 9\cdot 7)^{1/4} + \left(\frac{1}{5}\cdot 1\cdot 7\cdot 3\right)^{1/4} + \left(\frac{1}{9}\cdot\frac{1}{7}\cdot 1\cdot\frac{1}{3}\right)^{1/4} + \left(\frac{1}{7}\cdot\frac{1}{3}\cdot 3\cdot 1\right)^{1/4}\right\}} = 0.04 \quad (6.43)$$

$$w_4 = \frac{\left(\frac{1}{7}\cdot\frac{1}{3}\cdot 3\cdot 1\right)^{1/4}}{\left\{(1\cdot 5\cdot 9\cdot 7)^{1/4} + \left(\frac{1}{5}\cdot 1\cdot 7\cdot 3\right)^{1/4} + \left(\frac{1}{9}\cdot\frac{1}{7}\cdot 1\cdot\frac{1}{3}\right)^{1/4} + \left(\frac{1}{7}\cdot\frac{1}{3}\cdot 3\cdot 1\right)^{1/4}\right\}} = 0.09 \quad (6.44)$$

Note that the sum of the weighting factors of (6.41, 6.42, 6.43 and 6.44), $\sum_{i=1}^{n} w_i$ equals 1:

$$\sum_{i=1}^{4} w_i = 0.65 + 0.22 + 0.04 + 0.09 = 1 \quad (6.45)$$

The computed priority vector $W = (0.65, 0.22, 0.04, 0.09)^T$, using the geometric mean method is identical to the one in (6.32) computed using the eigenvector method. The same method can be applied to all the other comparison matrices discussed in Sect. 6.3.1.

4 The Approximation Method

Applying the eigenvector method can be rather cumbersome. The geometric mean is much easier to apply, but it does not determine the principal eigenvalues. For smaller size comparison matrices A, an approximation method can be used. In general, this method works quite well up to 4x4 comparison matrices A.

We demonstrate this method for determining the approximated priority vector $\widetilde{W}$ and the principal eigenvalue $\widetilde{\lambda}_{max}$ of comparison matrix A of (6.27) expressing the pairwise comparison of the four decision criteria SA, LE, JE, and JS.

Using this method, the sum of the values per column of comparison matrix A has to be computed first:

$$A = \begin{bmatrix} 1 & 5 & 9 & 7 \\ \frac{1}{5} & 1 & 7 & 3 \\ \frac{1}{9} & \frac{1}{7} & 1 & \frac{1}{3} \\ \frac{1}{7} & \frac{1}{3} & 3 & 1 \end{bmatrix} \tag{6.46}$$

$$\text{Sum} \quad \frac{13}{9} \quad \frac{13}{2} \quad 20 \quad \frac{34}{3}$$

Next, the element values of matrix A of (6.27) are normalized by dividing the elements in the column of comparison matrix A by the sum of the element values for each column, as expressed in (6.46). This yields the *normalized approximated comparison matrix* $\widetilde{A}$ of which the sum of the elements of each column is 1.

$$\widetilde{A} = \begin{bmatrix} \frac{9}{13} & \frac{10}{13} & \frac{9}{20} & \frac{21}{34} \\ \frac{1}{5}\cdot\frac{9}{13} & \frac{2}{13} & \frac{7}{20} & \frac{9}{34} \\ \frac{1}{9}\cdot\frac{9}{13} & \frac{1}{7}\cdot\frac{2}{13} & \frac{1}{20} & \frac{1}{34} \\ \frac{1}{7}\cdot\frac{9}{13} & \frac{1}{3}\cdot\frac{2}{13} & \frac{3}{20} & \frac{3}{34} \end{bmatrix} \tag{6.47}$$

Finally, the elements of each row in (6.47) are added up and the sum of each row is divided by the number of elements in this row to obtain the *approximated priority vector* $\widetilde{W}$ as expressed in (6.48).

$$\widetilde{W} = \frac{1}{4}\cdot\begin{bmatrix} 9/13 + 10/13 + 9/20 + 21/34 \\ 1/5\cdot 9/13 + 2/13 + 7/20 + 9/34 \\ 1/9\cdot 9/13 + 1/7\cdot 2/13 + 1/20 + 1/34 \\ 1/7\cdot 9/13 + 1/3\cdot 2/13 + 3/20 + 3/34 \end{bmatrix} = \begin{bmatrix} 0.63 \\ 0.23 \\ 0.04 \\ 0.10 \end{bmatrix} \tag{6.48}$$

The resulting approximated priority vector $\widetilde{W} = (0.63, 0.23, 0.04, 0.10)^T$ is quite similar to the exact computed priority vector $W = (0.65, 0.22, 0.04, 0.09)^T$ of (6.32).

The *approximated principal eigenvalue* $\widetilde{\lambda}_{max}$ is computed by multiplying each element of the approximated priority vector $\widetilde{W}$ of (6.48) with the corresponding sum of each column of comparison matrix A of (6.46) for the corresponding elements. This results in:

$$\widetilde{\lambda}_{max} = \left(0.63 \cdot \frac{13}{9}\right) + \left(0.23 \cdot \frac{13}{2}\right) + (0.04 \cdot 20) + \left(0.10 \cdot \frac{34}{3}\right) = 4.375 \quad (6.49)$$

The approximated principal eigenvalue $\widetilde{\lambda}_{max} = 4.375$ differs slightly from the exact calculated eigenvalue $\lambda_{max} = 4.206$ in Sect. 6.3. The approximation method can be further applied to all the other comparison matrices of the Green Waste Valorization case discussed in Sect. 6.3.

5 AHP Using Excel

MS Excel can be very helpful in solving AHP cases. In this section, we demonstrate how to derive the priority vector in Excel using the geometric mean method and the approximation method applied to the Green Waste Valorization case.

First, the comparison matrices have to be put in an Excel sheet. It is convenient to select the number format "*Fraction*" when entering the input variables of a comparison matrix. Only the cells of the upper triangular comparison matrix, colored in grey in Fig. 6.7, have to be entered (for black & white copies, these are the cells B16-E16; C17-E17; D18-18 and E19). The values belonging to the lower triangular comparison matrix, colored in orange, are calculated as the reciprocal values of the upper triangular matrix.

The values of the lower triangular comparison matrix are calculated as the reciprocal values of the upper triangular matrix, as depicted in Fig. 6.8 for the cells B17:19, C18:19, and D19. The sum of each column is calculated in B21:E21 as discussed for (6.46):

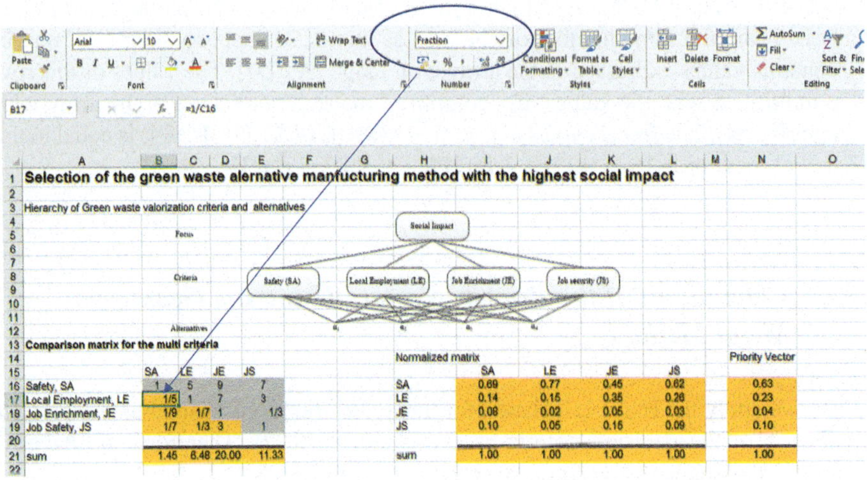

Fig. 6.7 Excel comparison matrix for the criteria of the Green Waste Valorization case

	A	B	C	D	E
15		SA	LE	JE	JS
16	Safety, SA	1	5	9	7
17	Local Employment, LE	=1/C16	1	7	3
18	Job Enrichment, JE	=1/D16	=1/D17	1	0.3333333333333
19	Job Safety, JS	=1/E16	=1/E17	=1/E18	1
20					
21	sum	=SUM(B16:B19)	=SUM(C16:C19)	=SUM(D16:D19)	=SUM(E16:E19)

Fig. 6.8 Excel formulas for the reciprocal values in the approximated comparison matrix of the decision criteria of the Green Waste Valorization case

The same goes for all the other comparison matrices discussed in this chapter. We will first discuss how to construct the Excel formulas for calculating priority vectors using the geometric mean method. This is the easiest way to deal with the AHP using Excel.

In Sects. 6.5.1 and 6.5.2, the following MS Excel functions are used:

PRODUCT (array): The PRODUCT function in Excel returns the product of numbers provided as arguments expressed in the array or range [R_1: R_n]. R_1 and R_n refer to the start and end cells in the Excel spreadsheet of array [R_1:R_n]. The expression A1 = PRODUCT(A10:A14) is equal to the expression A1 = A10*A11*A12*A13*A14.

POWER (number, power): The POWER function in Excel returns a number to a given power. The POWER function works like an exponent in a standard math equation. The expression A1 = POWER (2,3) is equal to the expression A1 = 2^3, which corresponds to the mathematical notation of 2^3.

MMULT (array 1, array 2): The MMULT function in Excel returns the matrix product of two arrays. Each array corresponds with a range of subsequent Excel cells. The function MMULT is used to calculate the matrix product of two matrices by multiplying the elements of each row of matrix 1 with the elements of each column of matrix 2. This results in a matrix that has the same number of rows as matrix 1 and the same number of columns as matrix 2. Suppose the row of matrix 1 contains values expressed in the array with range [A1:D1]. The values of this row are multiplied with the values of a column of matrix 2 expressed in the array with range [F1:F4]. The Excel expression A10 = MMULT (A1:D1, F1:F4) is equal to the Excel expression A10 = A1*F1 + B1*F2 + C1*F3 + D1*F4.

5.1 Calculation of the Priority Vectors Using the Geometric Mean Method

First, we apply the Excel calculation for the geometric mean method to comparison matrix A of the decision criteria as expressed in (6.27). Therefore, we have to compute the elements w_i of priority vector W using Excel.

The sum of the columns for each row i of matrix A is calculated in cells I16 to I19. This corresponds with the term $\prod_{j=1}^{n} a_{ij}$ in eq. (6.19). We use the Excel function PRODUCT to multiply all the elements of comparison matrix A belonging to a specific row i. As an example, the product of elements of the first row ($i = 1$) of comparison matrix A of (6.27) is:

$$\prod\nolimits_{j=1}^{4} a_{1j\cdot} = 1 \cdot 5 \cdot 9 \cdot 7 = 315. \tag{6.50}$$

This value is computed in cell I16 as I16 = PRODUCT(B16:E16) (see Figs. 6.9, 6.10 and 6.11).

Next, the computed values in the cell range [I16:I19] are powered to $1/n$ in the corresponding cell range J16:J19. Note that the rank n of matrix A is defined as a grey cell, meaning the value has to be inserted in Excel. In this case, the matrix A rank value n has been inserted in cell Q16 (see Fig. 6.10) and equals 4. The value of Q16 is 4 since A is a 4x4 matrix. The nth root, i.e., in this case the 4th root of 315 or $(315)^{1/4}$ is computed in cell J16 using the function POWER, J16 = POWER(I16;1/$Q$16) (see Fig. 6.11).

To calculate priority vector W using the geometric mean method, the elements of priority vector W are computed in the cell range N16:N19 following eq. (6.19).

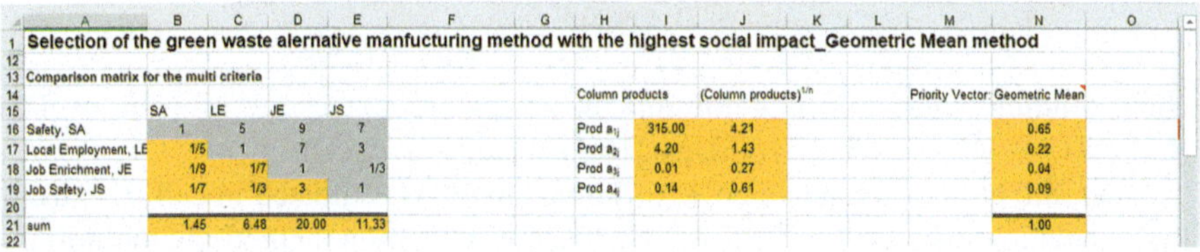

Fig. 6.9 Excel sheet organization to compute a priority vector using the geometric mean method

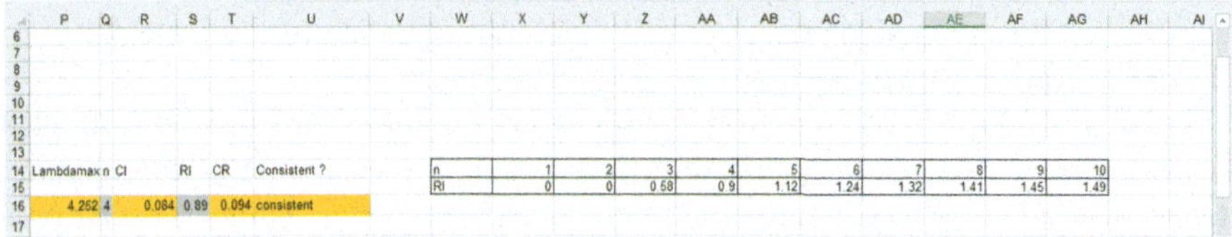

Fig. 6.10 Excel sheet organization to compute the consistency ratio CR for the in Fig. 6.9 depicted comparison matrix A

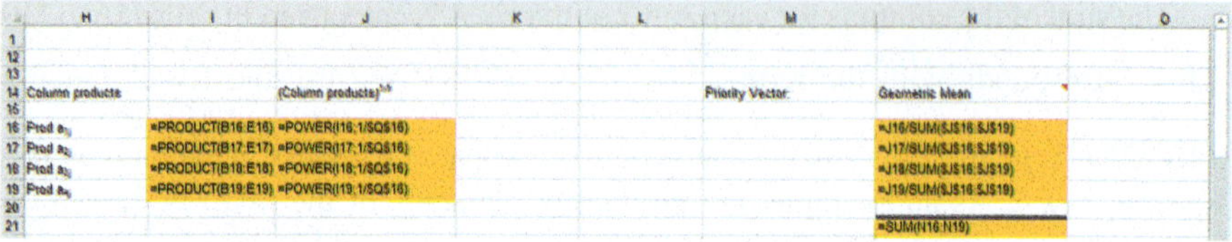

Fig. 6.11 Excel functions used to compute the priority vector depicted in Fig. 6.9 using the geometric mean method

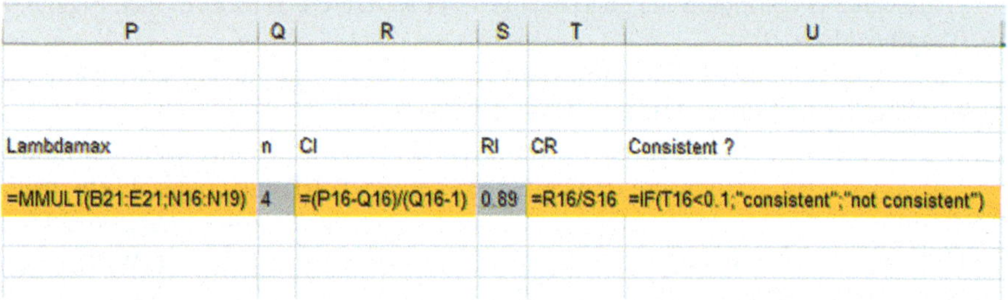

Fig. 6.12 Excel functions used to compute λ_{max}, CI and CR for the in Fig. 6.9 depicted comparison matrix A

Take $w_1 = \frac{(1\cdot5\cdot9\cdot7)^{\frac{1}{4}}}{\left\{(1\cdot5\cdot9\cdot7)^{\frac{1}{4}}+\left(\frac{1}{5}\cdot1\cdot7\cdot3\right)^{\frac{1}{4}}+\left(\frac{1}{9}\cdot\frac{1}{7}\cdot1\cdot\frac{1}{3}\right)^{\frac{1}{4}}+\left(\frac{1}{7}\cdot\frac{1}{3}\cdot3\cdot1\right)^{\frac{1}{4}}\right\}} = 0.65$. The corresponding values of this formula in Excel are expressed in N16 = J16/(J16 + J17 + J18 + J19) = J16/SUM(J16:J19) (see Fig. 6.11).

Next, the principal eigenvalue λ_{max} (Lambdamax) is estimated in cell P16, using the calculation discussed in Sect. 6.4 for the approximation method, by multiplying the sum of elements per column of matrix A displayed in range [B21:E21] with the corresponding values of priority vector W displayed in range [N16:N19]. In Excel, this is expressed as P16= =MMULT(B21:E21;N16:N19) (see Fig. 6.12).

The other priority vectors belonging to each pairwise comparison matrix representing the pairwise comparison of alternatives per criterion are calculated the same way. The comparison matrices and their corresponding priority vectors, λ_{max}, CI, and CR are depicted in Figs. 6.13, 6.14, 6.15, and 6.16.

Finally, the priority vectors of the pairwise comparison of the investement alternatives per criterion and the pairwise comparison of the social criteria are combined to calculate the resulting preference weighting factors (largest number = first preference) as illustrated in Figs. 6.17 and 6.18. The same priority ranking of the four alternatives is obtained as depicted in Table 6.6. The composite weight of the alternatives differs slightly from the ones shown in Table 6.6 because of rounding differences.

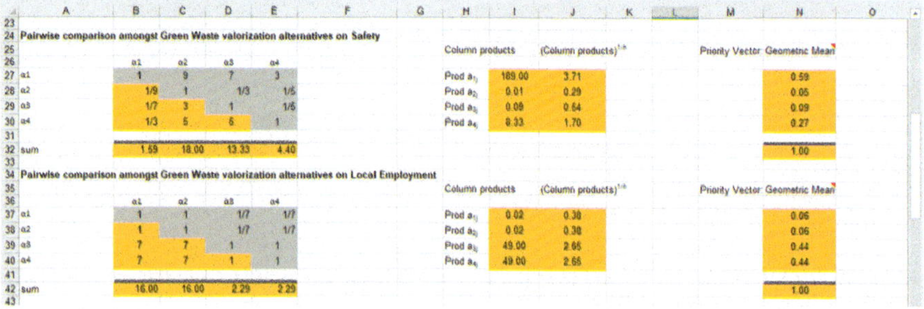

Fig. 6.13 Excel sheet organization to compute the comparison matrices A_{SA}, A_{LE} and priority vectors W_{SA}, W_{LE} using the geometric mean method

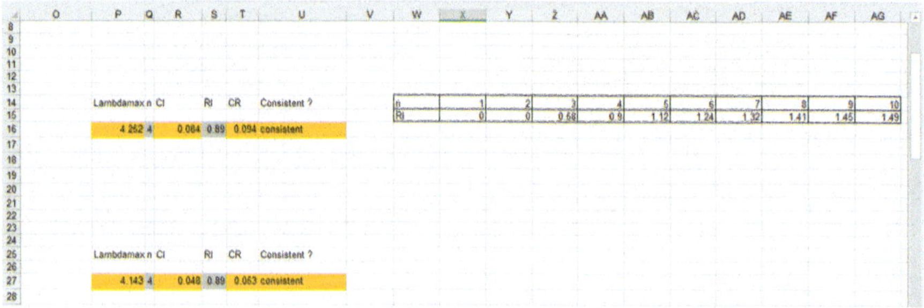

Fig. 6.14 Excel sheet organization to compute the calculation of λ_{max}, CI and CR for comparison matrices A_{SA} and A_{LE} using the geometric mean method

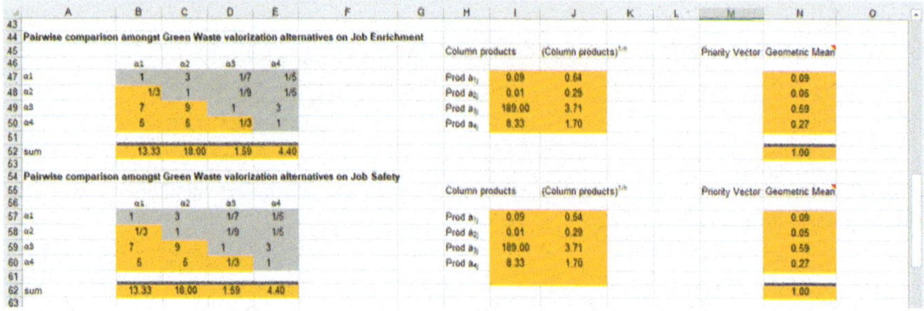

Fig. 6.15 Excel sheet organization to compute the comparison matrices A_{JE}, A_{JS} and priority vectors W_{JE}, W_{JS} using the geometric mean method

5.2 *Calculation of the Priority Vectors Using the Approximation Method*

Using the approximation method, the corresponding normalized approximated comparison matrix $\widetilde{A}$ (see (6.47) is calculated by dividing the value of each element in the

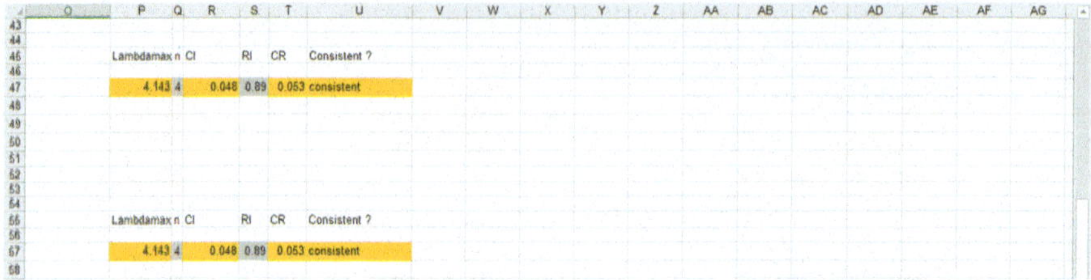

Fig. 6.16 Excel sheet organization to compute the calculation of λ_{max}, *CI* and *CR* for comparison matrices A_{JE} and A_{JS}

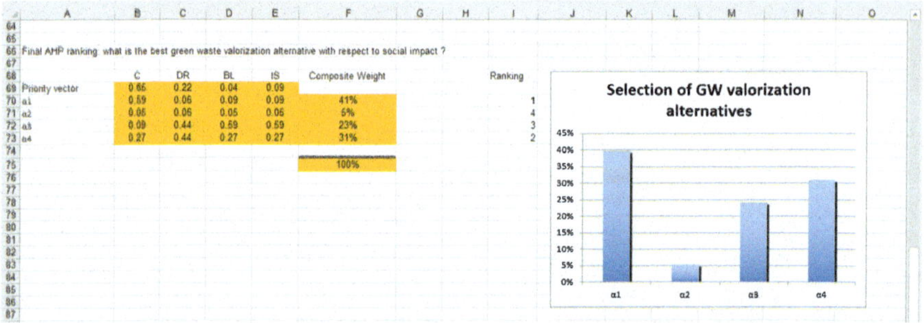

Fig. 6.17 Final ranking of the four investement alternatives using the geometric mean method

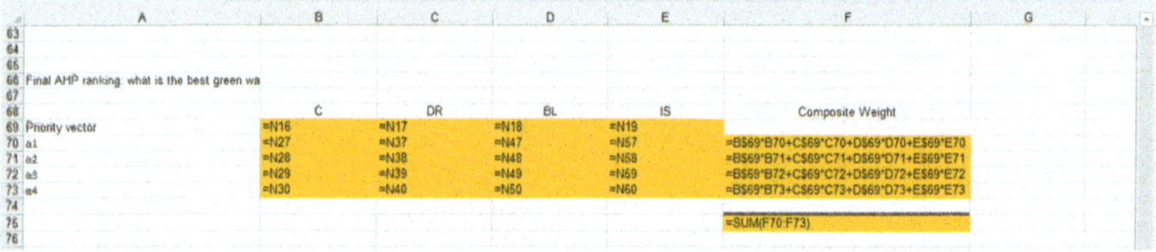

Fig. 6.18 Excel functions used to compute the ranking for the four investement alternatives using the geometric mean method

comparison matrix by the sum of the corresponding column. Next, the approximated priority vector $\widetilde{W}$ (see (6.48)) is calculated as the average of the row elements for each criterion. Comparison matrix *A* and the normalized approximated comparison matrix $\widetilde{A}$ for the multicriteria are depicted in Fig. 6.19, and the corresponding Excel formulas are shown in Fig. 6.20.

Next, the approximated principal eigenvalue λ^*_{max} and the *CR* value are calculated (Fig. 6.21):

The approximated principal eigenvalue λ^*_{max} is estimated as described in (6.49), *CI* and *CR* are calculated using eq. (6.17) and eq. (6.18), respectively. Only *n* for

	A	B	C	D	E	F	G	H	I	J	K	L	M	N	O
1	Selection of the green waste alernative manfucturing method with the highest social impact_Approximation method														
2															
12															
13	Comparison matrix for the multi criteria														
14								Normalized matrix						Priority Vector	
15		SA	LE	JE	JS				SA	LE	JE	JS			
16	Safety, SA	1	5	9	7			SA	0.69	0.77	0.45	0.62		0.63	
17	Local Employment, LE	1/5	1	7	3			LE	0.14	0.15	0.35	0.26		0.23	
18	Job Enrichment, JE	1/9	1/7	1	1/3			JE	0.08	0.02	0.05	0.03		0.04	
19	Job Safety, JS	1/7	1/3	3	1			JS	0.10	0.05	0.15	0.09		0.10	
20															
21	sum	1.45	6.48	20.00	11.33			sum	1.00	1.00	1.00	1.00		1.00	
22															

Fig. 6.19 Excel sheet organization to compute the normalized approximated comparison matrix $\widetilde{A}$ and the approximated priority vector $\widetilde{W}$

	H	I	J	K	L	M	N
1							
2							
12							
13							
14	Normalized matrix						Priority Vector
15		SA	LE	JE	JS		
16	SA	=B16/B$21	=C16/C$21	=D16/D$21	=E16/E$21		=AVERAGE(I16:L16)
17	LE	=B17/B$21	=C17/C$21	=D17/D$21	=E17/E$21		=AVERAGE(I17:L17)
18	JE	=B18/B$21	=C18/C$21	=D18/D$21	=E18/E$21		=AVERAGE(I18:L18)
19	JS	=B19/B$21	=C19/C$21	=D19/D$21	=E19/E$21		=AVERAGE(I19:L19)
20							
21	sum	=SUM(I16:I19)	=SUM(J16:J19)	=SUM(K16:K19)	=SUM(L16:L19)		=SUM(N16:N19)
22							

Fig. 6.20 Excel formulas used to compute the normalized approximated comparison matrix $\widetilde{A}$ and the approximated priority vector $\widetilde{W}$

	P	Q	R	S	T	U	V
1							
2							
12							
13							
14	Lambdamax	n	CI	RI	CR	Consistent ?	
15							
16	4.375	4	0.125	1	0.141	not consistent	
17							

	W	X	Y	Z	AA	AB	AC	AD	AE	AF	AG	AH
14	n	1	2	3	4	5	6	7	8	9	10	
15	RI	0	0	0.58	0.9	1.12	1.24	1.32	1.41	1.45	1.49	

Fig. 6.21 Excel sheet organization to compute the *CR* value and the principal eigenvalue λ^*_{max}

eq. (6.17) and *RI* have to be entered in the Excel spreadsheet as input variables (Fig. 6.22):

Likewise, the pairwise comparison matrices, priority vectors, and consistency checks have to be inserted for the pairwise comparison of the four alternatives for each of the four criteria (Figs. 6.23 and 6.24).

Finally, the priority of the four alternatives is calculated, taking the pairwise comparison of the alternatives and criteria into account (Fig. 6.25).

The formulas used are depicted in Fig. 6.26.

The final ranking shown in Fig. 6.25 is quite similar to the one derived using the geometric mean method, as depicted in Fig. 6.17. Note that the priority vectors only differ slightly between those calculated with the geometric mean method and those calculated with the approximation method in this example.

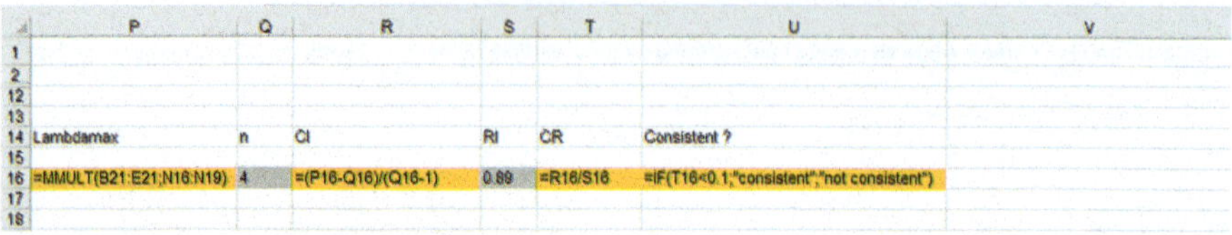

Fig. 6.22 Excel formulas for the computation of the *CR* value and the principal eigenvalue λ^*_{max}

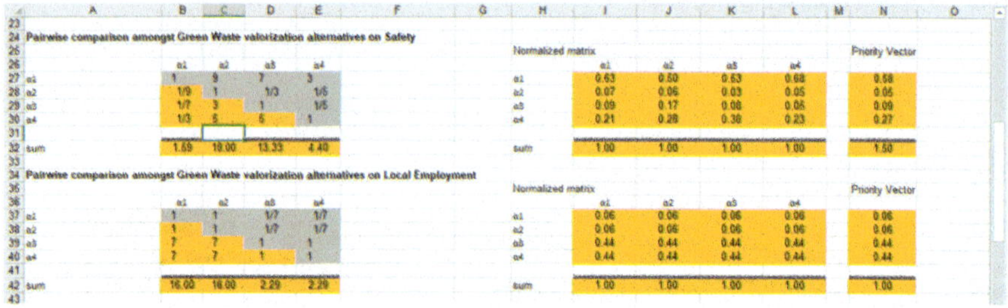

Fig. 6.23 Excel sheet organization to compute the pairwise comparison of alternatives for SA and LE of the Green Waste Valorization case using the approximation method

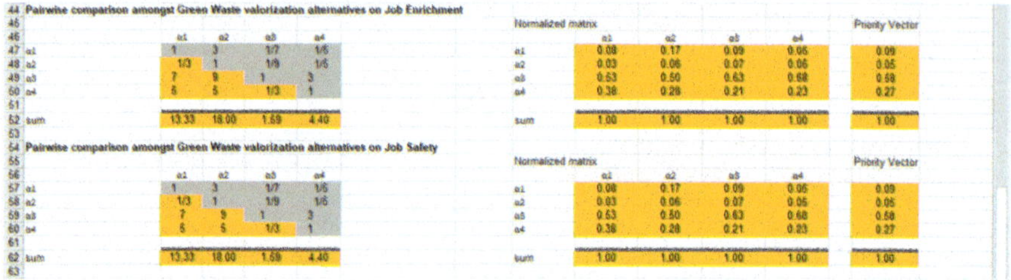

Fig. 6.24 Excel sheet organization to compute the pairwise comparison of alternatives for JE and JS of the Green Waste Valorization case using the approximation method

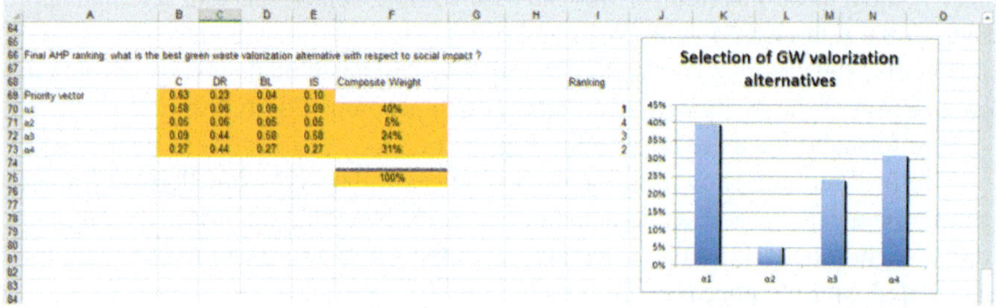

Fig. 6.25 Final ranking of the four investment options using the approximation method

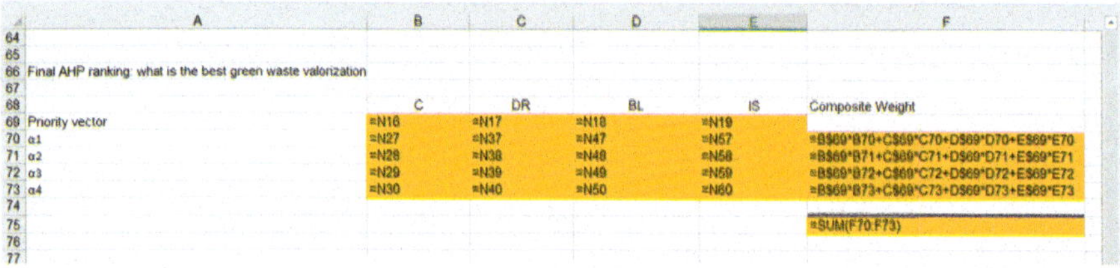

	A	B	C	D	E	F
64						
65						
66	Final AHP ranking: what is the best green waste valorization					
67						
68		C	DR	BL	IS	Composite Weight
69	Priority vector	=N16	=N17	=N18	=N19	
70	α1	=N27	=N37	=N47	=N57	=B$69*B70+C$69*C70+D$69*D70+E$69*E70
71	α2	=N28	=N38	=N48	=N58	=B$69*B71+C$69*C71+D$69*D71+E$69*E71
72	α3	=N29	=N39	=N49	=N59	=B$69*B72+C$69*C72+D$69*D72+E$69*E72
73	α4	=N30	=N40	=N50	=N60	=B$69*B73+C$69*C73+D$69*D73+E$69*E73
74						
75						=SUM(F70:F73)
76						
77						

Fig. 6.26 Excel formulas used to obtain the outcome in Fig. 6.25

Chapter 7
Systems Thinking and Introduction to System Dynamics Modeling

The Blind Men and the Elephant
John Godfrey Saxe (1816–1887)

It was six men of Indostan
To learning much inclined,
Who went to see the Elephant
(Though all of them were blind),
That each by observation
Might satisfy his mind.

The First approached the Elephant,
And happening to fall
Against his broad and sturdy side,
At once began to bawl:
"God bless me! but the Elephant
Is very like a WALL!"

The Second, feeling of the tusk,
Cried, "Ho, what have we here,
So very round and smooth and sharp?
To me 'tis mighty clear
This wonder of an Elephant
Is very like a SPEAR!"

The Third approached the animal,
And happening to take
The squirming trunk within his hands,
Thus boldly up and spake:
"I see," quoth he, "the Elephant
Is very like a SNAKE!"

D. Inghels, *Introduction to Modeling Sustainable Development in Business Processes*, https://doi.org/10.1007/978-3-030-58422-1_7

The Fourth reached out an eager hand,
And felt about the knee
"What most this wondrous beast is like
Is mighty plain," quote he:
"'Tis clear enough the Elephant
Is very like a TREE!"

The Fifth, who chanced to touch the ear,
Said: "E'en the blindest man
Can tell what this resembles most;
Deny the fact who can,
This marvel of an Elephant
Is very like a FAN!"

The Sixth no sooner had begun
About the beast to grope,
Then seizing on the swinging tail
That fell within his scope,
"I see," quoth he, "the Elephant
Is very like a ROPE!"

And so, these men of Indostan
Disputed loud and long,
Each in his own opinion
Exceeding stiff and strong,
Though each was partly in the right,
And all were in the wrong!

This poem by Saxe teaches us that different observation approaches can lead to different conclusions. Moreover, this poem teaches us that we are all somewhat blind in our observations: we often observe things the way we want to see the world. In general, we make use of a set of distinct thought patterns, including theories and research methods, for what is considered to be a legitimate way of approaching reality. These thought patterns are called *paradigms*. A famous example of a paradigm is how we have regarded the position of the earth in the universe over time. Initially, Ptolemy considered the earth as the center of the universe until Copernicus discovered that the earth turns around the sun.

Kuhn (1962) has shown that every significant breakthrough in science arises from a break with previous ways of thinking. He called these transitions *paradigm shifts*. A paradigm shift is about seeing the same observation in a radically different way.

With respect to sustainable development, two currently applicable paradigms need to be challenged. The first one is about how to model the complex world we live in. The Cartesian-Newtonian way of thinking (after René Descartes and Isaac Newton) is still dominant. Cartesian-Newtonian thinking, also known as analytical thinking, is based on the premise that a complex problem can be reduced to a set of separate smaller problems. By understanding this set of smaller problems, we can understand the complex problem.

However, Cartesian-Newtonian thinking has some drawbacks. By breaking up complex problems into a set of smaller, more manageable problems, the interaction between the parts gets lost. Moreover, Cartesian-Newtonian thinking is, in general, not very suitable when dealing with nonlinearities, which are a common feature of many real-life environmental and socio-economic problems. Chapters 2, 3, 4, 5, and 6 of this book dealt with solving sustainability problems in a traditional Cartesian-Newtonian way.

The second paradigm to be challenged is called the *paradigm of economic growth*, a term introduced by the ecological economist Herman Daly (1972) to characterize the belief in unlimited growth by mainstream economists.

1 Systems Thinking: Challenging the Cartesian-Newtonian Paradigm

Another approach to analyze sustainable development problems is systems thinking. Systems thinking, i.e., the process of thinking using system ideas, starts by observing how systems interact with each other. A system is a set of interconnected elements that form a whole; they show properties that are properties of the whole rather than of its components (Checkland 1981). Systems thinking emerged as a meta-discipline and meta-language that can be used to talk about the subject matter of many different fields (Currie and Galliers 1999). This way of thinking originated in biology in the first half of the twentieth century from the observation that analytical thinking was limited when attempting to understand how complex systems work as a whole. The discussion about vitalism, i.e., living things are much more than a sum of their parts, fueled the discussion that classic reductionist thinking, also known as analytical thinking, was not satisfactory when attempting to explain complex living structures in older disciplines such as physics and chemistry. In the late 1940s, the Austrian organismic biologist Ludwig von Bertalanffy founded the systems movement by arguing that ideas about organisms could be extended to complex wholes of any kind, i.e., to "systems" (von Bertalanffy 1968). The expectation of the systems thinkers of the 1940s and 1950s was that the scientific method would one day have two components: analytical thinking and systems thinking. Models based on system thinking are systems theoretic behavior models. In general, they are comprised of feedback loops between the different parts of a system and can easily deal with nonlinearities. These models are also holistic and allow synthesizing without a reduction of information. In systems models, the interactions are as important as the systems themselves. System thinking is very suited to modeling real-life, complex problems such as sustainable development problems because it supports bringing all kinds of often non-linear behavior together with economic, environmental, and societal behavior. Table 7.1 summarizes the main differences between analytical and system thinking.

Table 7.1 Main differences between analytical and systems thinking

Analytical (Cartesian-Newtonian) thinking	Systems thinking
Isolates elements and focuses on these elements	Starts from the whole of a system and focuses on the interactions between the elements of the system
Investigates the nature of interaction	Investigates the effects of interaction
Works with exact data and a precise approach	Employs a holistic approach
Independent of time	Considers time effects

Systems thinking interprets the universe as a series of interconnected and interrelated wholes rather than as linear cause-and-effect chains. It is a way of identifying the inherent organization in a complex situation, which has been called organized complexity. Systems thinkers make a distinction between dynamic complexity (the relationships between things) and detail complexity (details about things). It is an approach, a set of general principles and specific tools and techniques, rather than a subject area in its own right. Systems thinking can be applied to many different fields and is therefore described as a meta-discipline.

It is worth noting that Aristotle's concept that the whole is greater than the sum of its parts prevailed before the seventeenth century. Because of the great work of scholars such as Newton and Descartes, the dominant way of thinking shifted to analytical thinking in the seventeenth century.

Systems thinking has become a crucial component of complexity theory: we currently know that some complex systems exhibit dynamic behavior that can be extremely sensitive to initial starting conditions. This puts limits on the beliefs of the early systems thinkers that the dynamics of a system could be analyzed completely. However, this does not mean that systems thinking is no longer useful or that it has been superseded by complexity theory.

2 The Paradigm of Economic Growth

The belief in unlimited growth emerged amongst economists in the 1950s.

The almost continuous economic growth since the start of the Industrial Revolution about 200 years ago, measured in gross domestic product (GDP), is unprecedented when it is framed in a broader historical perspective. Before the 1820s, economic activity remained more or less unchanged year after year. Primarily due to a slow increase in the population, economic growth was about 0.05% per annum, in as far as it is possible to measure this retrospectively (Schmelzer 2015). This topic was also discussed in the first chapter.

It is remarkable to note that classic economists like Adam Smith, David Ricardo, and Thomas Malthus agreed that perpetually rising economic growth is not possible. They all agreed that a stationary state would ultimately develop. For Smith, this 'steady state' is characterized by a zero-population growth and a zero-capital accumulation (Ucak 2015). Economic growth became a self-evident concept in industrialized countries after World War II. This has been observed by Schmelzer (2015) by counting the number of scientific papers published in all academic journals in the JSTOR database that contain the term "economic growth", by discipline, 1930–2010 (see Fig. 7.1). However, we currently face environmental limits on absorbing ever-increasing levels of greenhouse gas emissions, and we are confronted with the fact that further economic growth in the developed world no longer improves health and happiness. In other words, economic growth no longer contributes to a growth in wellbeing in industrialized countries (Wilkinson and Pickett 2011). Further economic growth is also under pressure due to resource and energy scarcities, and internal structural problems such as over-accumulation and financialization, which lead to declining or stagnating growth rates. This could lead to growing income inequalities, and wealth increases as wages tend to grow much slower than returns on capital in times of slow economic growth (Piketty 2014).

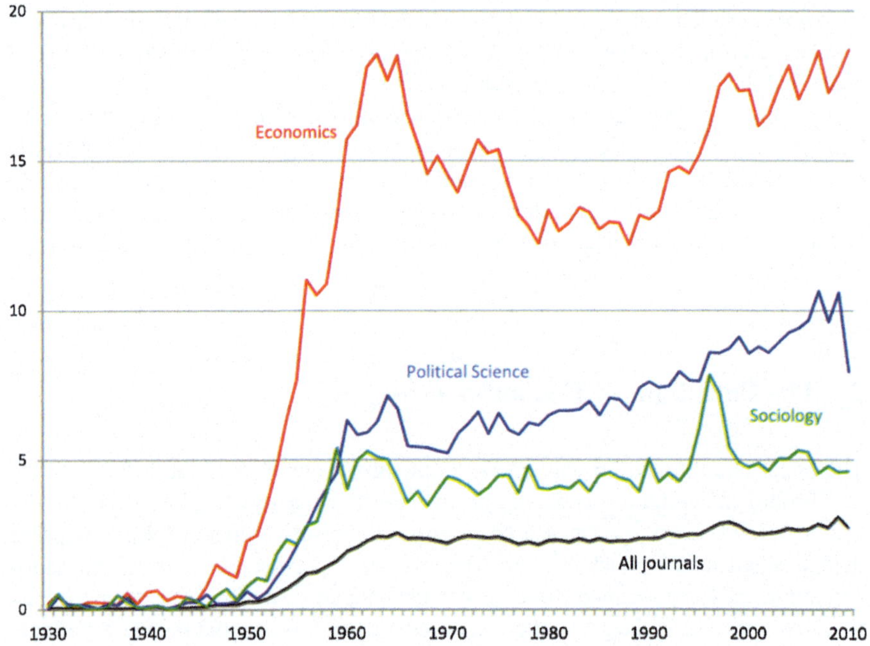

Fig. 7.1 Percentage of articles published in all academic journals in the JSTOR database that contain the term "economic growth", by discipline, 1930–2010. (Source: Schmelzer 2015)

3 Introduction to System Dynamics Modeling

Systems thinking can be practiced in many ways, including systems engineering and analysis, operations research, organizational cybernetics, interactive planning, organizational learning, and system dynamics.

System dynamics is concerned with building computer models of complex problem situations and then experimenting with and studying the behavior of these models over time. Such models often demonstrate how causal relationships, dynamic complexity, and structural delays may lead to counter-intuitive outcomes to improve a situation. System dynamic models make room for soft factors, such as motivation and perceptions, so that problem spaces can ultimately be better understood and managed (Caulfield and Maj 2001).

Jay Forrester of the Massachusetts Institute of Technology created system dynamics in the mid-1950s. In his banquet talk at the international meeting of the System Dynamics Society in Stuttgart, he described how system dynamics originated (Forrester 1989).

Jay Forrester was asked to lead the newly founded MIT Digital Computer Laboratory in 1947. His first task was the creation of MIT's first general-purpose digital computer. Nine years later, in 1956, Forrester accepted a professorship at the newly formed MIT School of Management.

Using hand calculations to examine the stock and flow feedback structure of a General Electric (GE) plant, Forrester showed that the instability of GE employment was not caused by externalities such as a business cycle, but by the corporate decision-making structure related to hiring and layoffs. This conclusion, based on insights about the stock and flow feedback structures that underlie electrical engineering, led to the creation of system dynamics. In 1961 he published his insights in the book "Industrial Dynamics," which is still a classic book in the field of system dynamics.

In the early stage of system dynamics, business and managerial problems were targeted almost exclusively. The link with the sustainable development domain came in 1971. In that year, the Club of Rome invited Jay Forrester to a meeting in Bern, Switzerland. The Club of Rome is an organization that is concerned about global crises that may appear in the future due to a surpassing of the earth's carrying capacity fueled by exponentially growing global populations. Forrester created the initial draft of a system dynamics model of the world's socio-economic system called WORLD1. The refined version of this model, WORLD2, was published in a book entitled World Dynamics. The WORLD2 model mapped important interrelationships between world population, industrial production, pollution, resources, and food. The model showed a collapse of the global socio-economic system sometime in the twenty-first century if no steps were taken to lessen the demands on the earth's carrying capacity. The model was also used to identify policy changes capable of moving the global system to a fairly high-quality state that is sustainable far into the future.

After the publication of World Dynamics, the Club of Rome funded an extensive study that resulted in the book "The Limits to Growth", published in 1972, which was written by Donella H. Meadows, Dennis L. Meadows, Jørgen Randers, and William W. Behrens III, who represented a team of 17 researchers (see reference: Meadows et al. 1972). The findings of this book are based on the WORLD3 model, which reemphasizes the earlier results of Jay Forrester's WORLD2 model. Almost 20 years later, in 1991, the authors redid the study, and the results were still consistent with the conclusions formulated in World Dynamics and The Limits to Growth.

Chapter 8
Causal Loop Diagrams

System dynamics looks at how various elements interact within a system over time and captures the dynamic aspects by incorporating concepts such as stock, flows, feedback, and delays. It provides an insight into the dynamic behavior of a system over time (Tang and Vijay 2001).

In this chapter, we will discuss how Causal Loop Diagrams (CLDs) are used to depict the feedback structure of systems. In the next chapter, we will discuss the fundamental modes of behavior of dynamic systems using CLDs to represent their underlying structure.

1 Causal Loop Diagram Notation

CLDs represent the structure of systems and are useful to display hypotheses about the causes of dynamics, to capture the mental models of individuals or teams, and to communicate the important feedbacks that one believes are responsible for a problem (Sterman 2000).

A causal loop diagram consists of variables connected by a causal influence called the *causal link*. The causal link is represented by an arrow and is assigned a polarity that can either be positive (+) or negative (−). It indicates how the dependent variable changes when the independent variable changes.

Consider two variables, an independent variable *X* and a dependent variable *Y*. There are only two ways *X* can influence *Y*. If *X* behaves the same way as *Y* (e.g., *Y* increases when *X* increases or *Y* decreases when *X* decreases) or depicts accumulations (i.e., *X* adds to *Y*), then a positive polarity is assigned to the arrow (Fig. 8.1). When *X* and *Y* behave oppositely or depict reductions (i.e., *X* subtracts from *Y*), then a negative polarity is assigned (Fig. 8.2).

D. Inghels, *Introduction to Modeling Sustainable Development in Business Processes*, https://doi.org/10.1007/978-3-030-58422-1_8

Fig. 8.1 A positive relationship between independent variable X and dependent variable Y

Fig. 8.2 A negative relationship between independent variable X and dependent variable Y

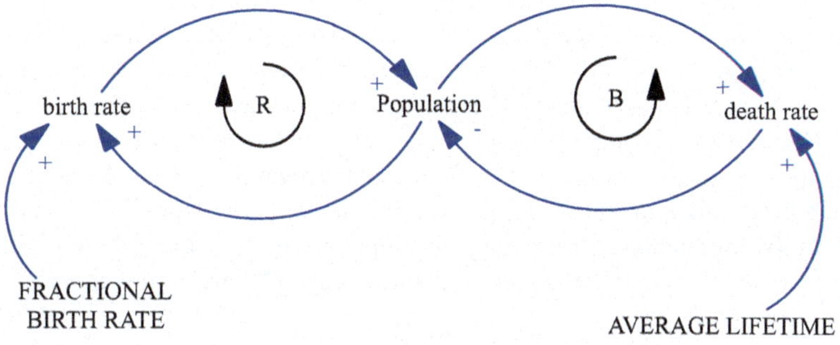

Fig. 8.3 An example causal loop diagram based on Sterman (2000)

An alternative convention that is sometimes used is to depict the + or – polarity by respectively the letters **s** (same) or **o** (opposite) to indicate the polarity of the causal link.

Causal links can form loops known as causal *feedback loops*. Feedback loops can be *positive* or *reinforcing*, denoted by a loop identifier in which the letter 'R' or the sign '+' in a clockwise arrow. Any situation where action produces a result that promotes more of the same action is representative of a reinforcing loop.

Feedback loops can also be *negative or balancing*, denoted by the letter 'B' or the sign '-' in a counterclockwise arrow. A balancing loop is one in which an action attempts to bring two things into agreement. Any situation where one attempts to solve a problem or achieve a goal or objective is representative of a balancing loop.

Figure 8.3 shows how to use the arrow notation convention discussed by Sterman (2000), which considers a population. The variable "Population" represents the number of people in the population. Births will increase the number of people in the population. Since the population will increase proportionally to the birth rate, they are connected by a positive causal loop. The birth rate itself depends on the population and the fractional birth rate. The latter is used to indicate how many new births occur on average in a given time frame per 100 people, for example. Therefore, the birth rate is the mathematical product of population and fractional birth rate. The more people, the higher the birth rate will be. The causal loop birth rate – population is a reinforcing feedback loop, indicated with the letter 'R' inside the reinforcing feedback loop.

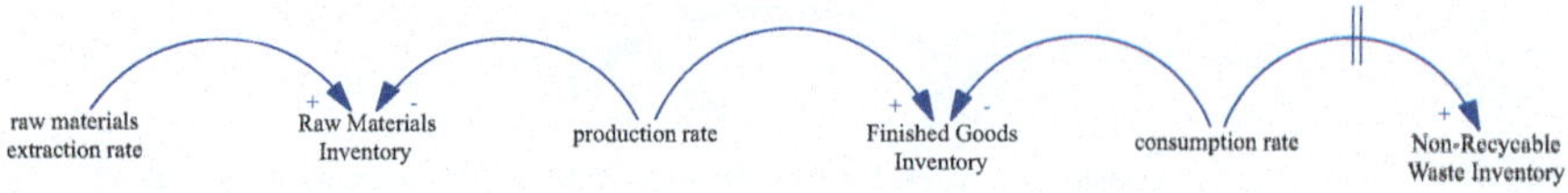

Fig. 8.4 Example of an open loop causal diagram

The number of people in the population will decrease because people eventually die after an average lifetime. The feedback loop population – death rate counteracts exponential population growth. This feedback loop is a balancing feedback loop, indicated with the letter 'B' inside the balancing feedback loop.

Another type of loop is the open loop. An open loop is a linear chain of causes and effects that does not close back on itself. Figure 8.4 represents an open loop causal diagram. It represents the linear economy, also known as "take-make-dump" economy. Contrary to the circular economy, there are no feedback loops. The raw materials inventory is defined as the difference between the raw materials extraction rate and the production rate. The finished goods inventory is defined as the difference between the production rate and the consumption rate. Consumed goods will, after some time, be discarded and added to the non-recyclable waste inventory.

2 Practical Use of Causal Loop Diagrams

2.1 *Avoid Spurious Correlations*

In a causal loop diagram (and the Stock and Flow model that will be discussed later), we only consider relationships that capture the underlying causal structure of the system under study. Confusing correlation with causality can lead to terrible misjudgments and policy errors (Sterman 2000). Before a correlation is depicted in a causal loop, one should check whether this correlation is not spurious. An example of a spurious correlation is depicted in Fig. 8.5. It is evident that the consumption of margarine is not correlated to the divorce rate in Maine. Other examples of spurious correlations are shown on: http://www.tylervigen.com/spurious-correlations

2.2 *Determining Loop Polarity*

Following Sterman (2000), there are two ways to determine whether the polarity of a causal loop is positive (reinforcing) or negative (balancing): a fast way and a safe way. The fast way consists of counting the number of negative causal links in the loop. If the number is even, the causal loop polarity is positive. If the number is odd, the causal loop polarity is negative. This is based on the property that causal loops showing negative polarity are self-correcting because they oppose disturbances.

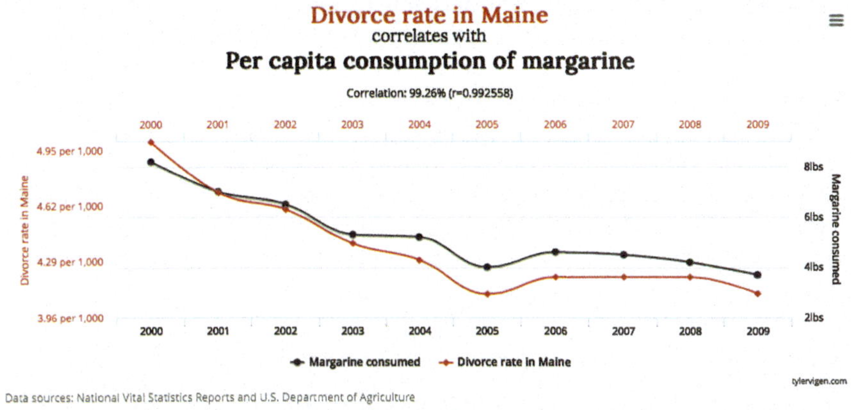

Fig. 8.5 An example of a spurious correlation between divorces and the consumption of margarine. (Source: http://www.tylervigen.com/spurious-correlations)

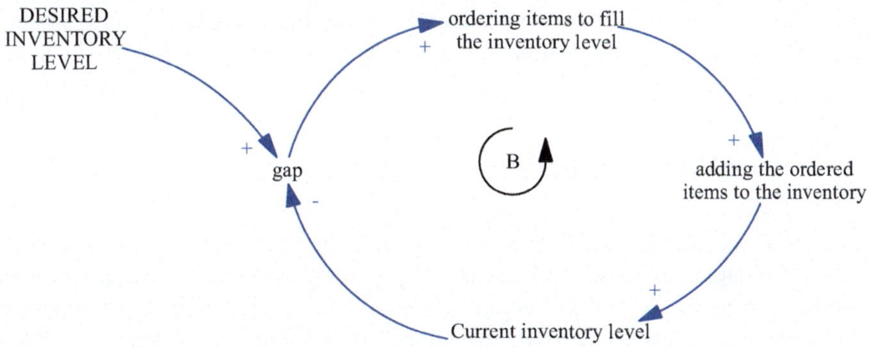

Fig. 8.6 Causal loop diagram depicting the inventory control of an item

However, one may miscount in complex diagrams, or one could be mistaken if a polarity link was mislabeled.

The safe way to determine the polarity of a causal loop is to examine the effect of a small change in one of the variables in the loop. If it affects a reinforcement of the small change, it is a positive feedback loop. Otherwise, it is a negative feedback loop.

The example of adding items to the inventory to keep the desired inventory level demonstrates how the causal loop polarity can be determined for a simple CLD (Fig. 8.6).

When the desired inventory level of an item is higher than the current inventory level, the gap or difference between the desired number of items and the current number is ordered for replenishment. By adding the ordered items to the inventory, the current inventory level will increase. If the amount that can be ordered is limited in time, the gap may become smaller after a delivery, but it will still be there. The process continues until the gap disappears.

The behavior of the variables in this simple control loop is expressed by the polarity on the causal links that connect these variables. For example, ordering items will lead to adding items to the inventory. Since both variables behave in the same way, a positive polarity is assigned. When the current inventory level increases due to adding the ordered items, the gap with the desired inventory level decreases. Therefore, a negative polarity is assigned to the variable gap.

All the different loops described form a closed loop. The sign for this particular simple closed loop is determined by counting the number of minus signs on all the links that make up the loop. The sign of the feedback loop is negative since there is only one negative causal link in this loop.

2.3 Depicting Important Delays in Causal Links

Significant delays create inertia in systems causing oscillation in the system's behavior. Therefore, causal flow diagrams should indicate these significant delays. Delays are indicated on the causal link with a double bar, as indicated in Fig. 8.7. We assume in this figure that the delivery time of ordering additional items is substantially high.

2.4 Variable Names in Causal Diagrams

The variables used in causal diagrams should be nouns or noun phrases such as "costs", "GHG emissions", "job security", etc. Make sure that the causal diagram can be read as the system's structure. Verbs cannot be used since they express a direction (e.g., "decreasing GHG emissions" is not allowed to denote a causal diagram variable because decrease involves a direction for the variable).

Interested readers are encouraged to visit: http://www.systems-thinking.org/intst/int.htm.

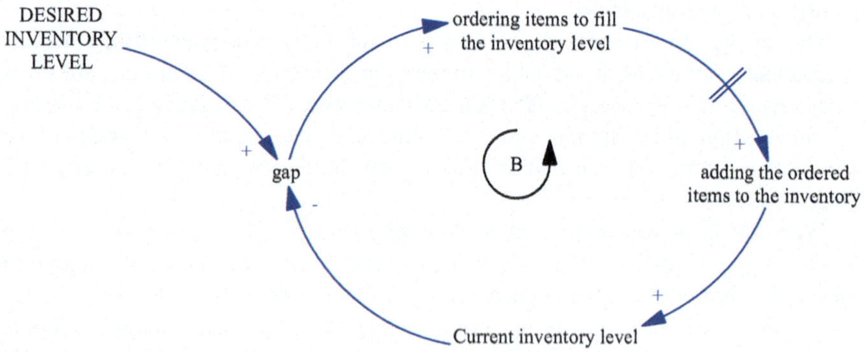

Fig. 8.7 Causal loop diagram depicting a substantial delay in delivering the ordered items

Chapter 9
Structure and Fundamental Modes of Behavior in Dynamic Systems

You may have already observed that many different systems behave the same way. For example, there is a similarity between global population growth over the last 200 years and the annual accumulation of money on a saving account given a fixed yearly net interest rate. Both systems show the same exponential growth behavior because they have a common underlying structure somewhere. The behavior of systems is generally determined by their structure. That structure may consist of feedback loops, stocks and flows, or nonlinearities created by the structure's interaction with the decision-making processes (Sterman 2000).

In this chapter, we discuss the three basic modes of behavior in dynamic systems and their underlying structure: exponential growth, goal seeking, and oscillation. Then we discuss the behavior of more complex dynamic systems and how they interact with some basic structures: S-shaped growth, S-shaped growth with overshoot, S-shaped growth with overshoot and collapse, and tragedy of the commons. These so-called archetypes of system behavior are often observed in the domain of sustainability, as we will show in the following sections.

1 Basic Modes of Dynamic Behavior

1.1 Exponential Growth

The underlying basic structure of *exponential growth* behavior is positive feedback, also known as reinforcing feedback. The state of an exponential growth system at a given time is the previous state of that system multiplied by a net increase rate, as depicted in the causal loop diagram in Fig. 9.1.

A typical example is the *compound interest* on a deposit account. Suppose you start with an initial deposit amount of €100 in year $t = 0$ and a net interest of 2% is guaranteed for all subsequent years. At the beginning of year $t = 1$, the net amount

D. Inghels, *Introduction to Modeling Sustainable Development in Business Processes*, https://doi.org/10.1007/978-3-030-58422-1_9

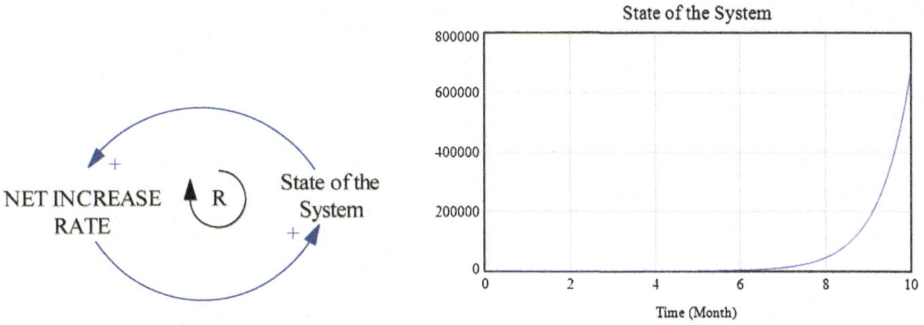

Fig. 9.1 State and structure of an exponential growth system – CLD and illustrative example

on your account will be 100 + 100·2% = €102. If the net interest does not change, the net amount on the account at the beginning of year $t = 2$ will be 102 + 102·2% = €104.04. Generalized, the amount on the account at year t, with a constant net interest of i% with an initial deposit amount A_0 will be $A_t = A_0 \cdot (1 + i)^t$ [€]. Compound interest is an example of pure exponential growth. The doubling time is constant. It takes the same amount of time, T_d, to double the amount on the account for an initial deposit of A_0 = €100 to the amount A_{Td} = €200 as for A_0 = €200 to A_{Td} = € 400. The doubling time T_d if the net yearly interest rate of 2% is constant is: $T_d = \frac{ln\,2}{0.02} \cong 35$ years.

There is a significant difference between linear and exponential growth. *Linear growth* is independent from the state of the system. Suppose you can get 2% fixed interest each year on the original amount of money A_0 on your deposit account. With linear growth, the yearly net increase will be €2 each year for A_0 = €100; this will not vary no matter what amount is on the deposit account. The amount on your account in year t will be $A_t = A_0 + t \cdot 2\% \cdot A_0$ [€].

Figure 9.2 shows the difference in behavior between linear and exponential growth applied to the example above. Please observe that both systems behave quite similarly for the first 10 years. After the doubling time T_d = 35 years, for exponential growth, the net amount on the account will have doubled to A_{35} = €200, whereas for linear growth, the amount will be €170 at t = 70 years.

After another doubling period of 35 years, at t = 70 years, the net amount for exponential growth will be double the amount at t = 35 years; in other words A_{70} = €400. For linear growth, the net amount will be €240.

Exponential growth is a behavior related to positive feedback in the underlying structure. If the net increase rate is positive, this will generate *reinforcing growth*. If the net increase rate had been negative, this would generate *reinforcing decline*.

There are no examples of indefinite exponential growth in real life. Most of the time, exponential growth is only a temporary state in a system whereby the net increase rate may differ over time. Let's take the global annual growth of electric vehicle sales as an example (see Fig. 9.3). Because a growing number of countries, regions, and cities are imposing increasingly more severe greenhouse gas emission restrictions on cars, the global electric car stock grew exponentially in the period

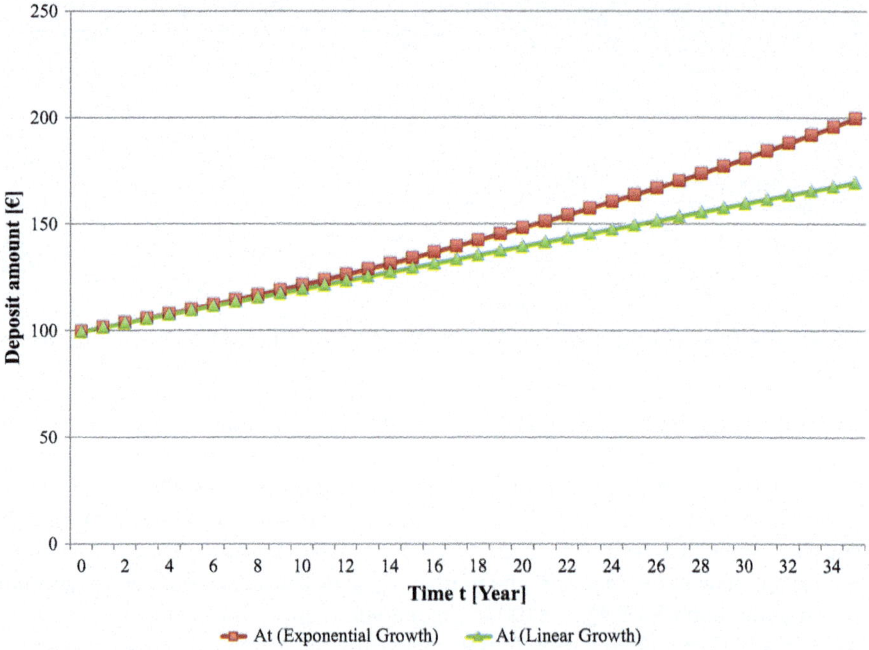

Fig. 9.2 Different behavior between exponential and linear growth applied to a saving account for the same interest rate of 2% and initial deposit of €100

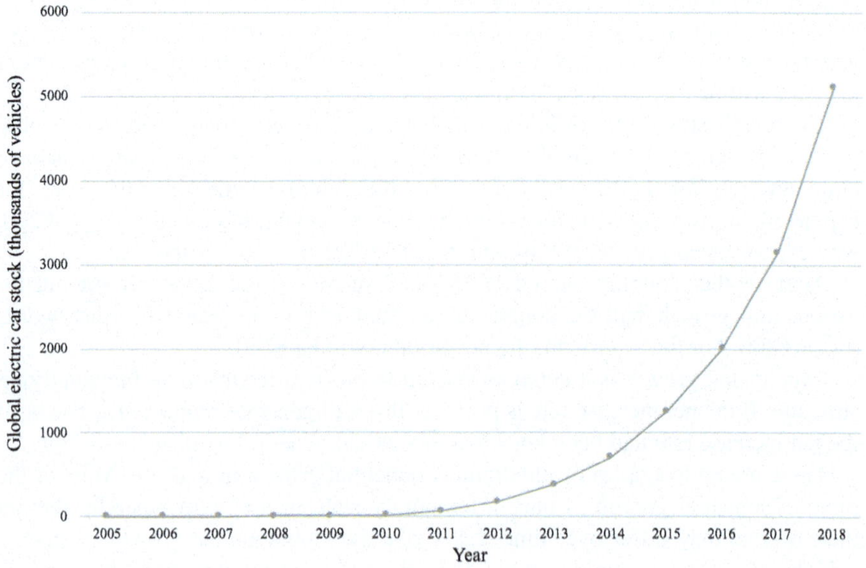

Fig. 9.3 Global electric car stock (battery electric cars [BEV] and plug-in hybrid cars [PHEV]) evolution between 2005 and 2018. (Source: IEA 2019)

2005–2018. Once the market is saturated, which may take many years, the exponential growth will gradually stop.

Other examples of exponential growth related to the domain of sustainable development include population growth in developed countries since the start of the Industrial Revolution (see Chap. 1), and the amount of greenhouse gas emissions related to the burning of fossil fuels over the last 200 years.

1.2 Goal Seeking

Systems demonstrating goal-seeking behavior finally demonstrate equilibrium, i.e., the state of the system no longer varies (see Fig. 9.4). The underlying structure is based on negative feedback, also known as balancing feedback. This negative feedback originates in a negative causal loop that includes a process to compare the desired and actual conditions. The desired condition of the system is the goal that one strives to attain. As long as the state of the system is different from the goal state, the gap (i.e., the difference between the current and the desired state) determines the rate at which the state of the system approaches its goal. This rate decreases when the gap with the desired state decreases.

Controlling the room temperature with a heating device is a typical example of a system displaying goal seeking behavior. Suppose the initial room temperature at $t = 0$ is $T_O = 15$ °C, and the desired (or goal) temperature is $T_g = 20$ °C. Initially, at $t = 0$, the gap between the desired and actual room temperature is $T_g - T_O = 5$ °C. This measured gap is the input for the thermostat, which heats the room proportionally to the measured gap. The larger the gap, the more power is put into the heating device. The room temperature starts to rise and becomes, at a given time t, $T_t = 19$ °C. The gap with the desired temperature is $T_g - T_t = 1$ °C. The thermostat will steer the heating device in such a way that the power drops. The power will be killed when the room temperature reaches the desired goal of $T_g = 20$ °C. This example will be further explored in Chap. 10.

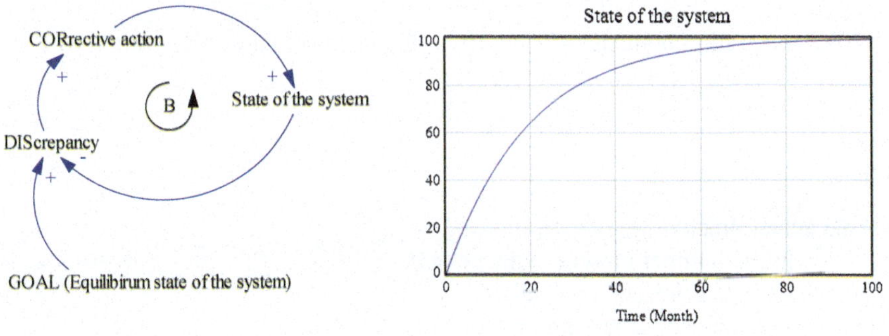

Fig. 9.4 State and structure of a goal seeking system – CLD and illustrative example (Goal = 100)

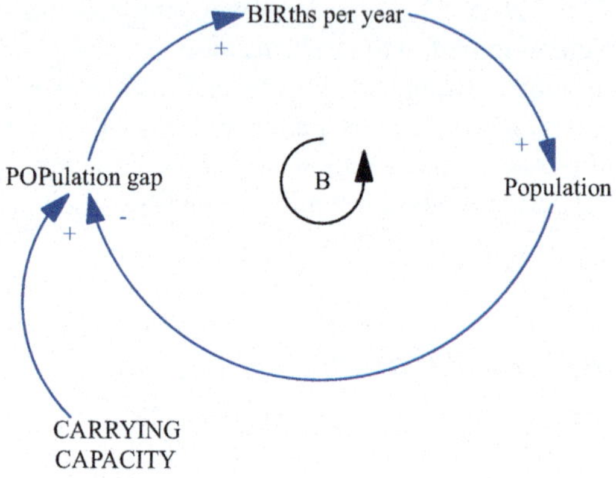

Fig. 9.5 Causal loop diagram of constrained population growth

An example of goal seeking behavior in the domain of sustainable development is constrained population growth. This occurs when there is a carrying capacity for the number of people in a population, and where a control mechanism is in place so that this number is not exceeded. As soon as the carrying capacity is reached, no births will occur until the population starts declining. The causal loop is depicted in Fig. 9.5.

A special case of goal seeking behavior is called *exponential decay*. Systems showing exponential decay behavior have the remarkable property that there is a linear relationship between the size of the gap and the adjustment rate. As the gap falls, so does the adjustment rate. Exponential decay is characterized by its *half time*. This is the time it takes for half of the remaining gap to be eliminated.

A quantity is subject to exponential decay if it decreases at a rate proportional to its current value. Exponential decay decrease starts fast and becomes slower. This is shown in the problem with radioactive waste. Remember from the previous section that true exponential decrease occurs whenever the rate of change of something is proportional to the thing itself. The rate of change of decay is presented as:

$$\frac{dN}{dt} = -d \cdot N(t) \tag{9.1}$$

After integrating both sides of Eq. (9.1), the exponential decay formula is:

$$N(t) = N_0 \cdot e^{-d \cdot t} \tag{9.2}$$

Where:

t is the actual time,
N_o is the initial number present at time $t = 0$,
$N(t)$ is the number present after time t,
d is the decay rate constant with unit $1/t$ or t^{-1}.

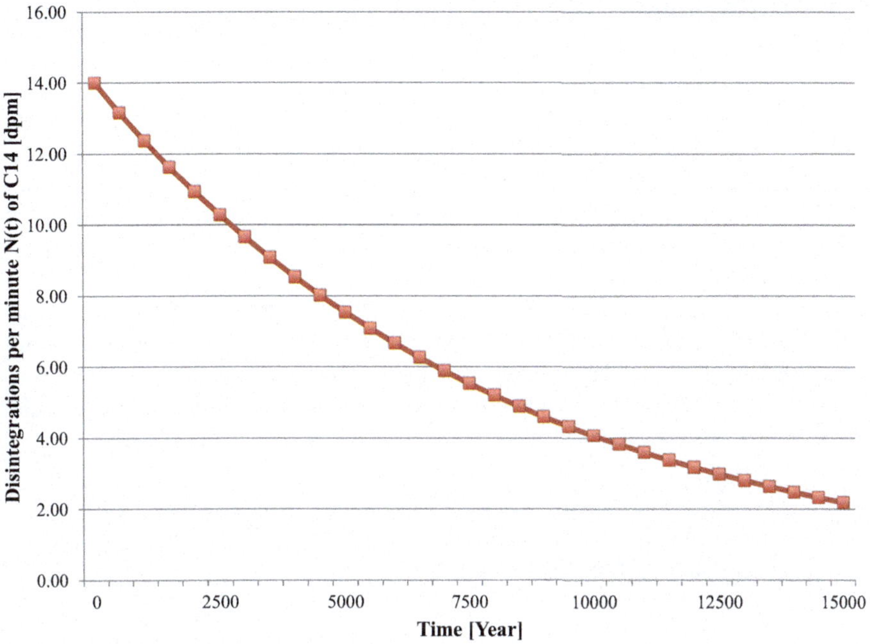

Fig. 9.6 Radioactive exponential decay behavior for C_{14}

Instead of using the decay constant, the reciprocal *Adjustment Time* (*AT*) or the time constant is often used:

$$AT = \frac{1}{d} \tag{9.3}$$

Given a half-life time $T_{1/2}$, the decay rate d is presented as:

$$d = \frac{\ln(2)}{T_{1/2}} \tag{9.4}$$

The exponential decay formula is used to calculate the number of atoms remaining during the radioactive decay of any nucleus, amongst other things. Take, as an example, the famous C_{14} (Carbon fourteen), which is used to find the approximate age of objects. The radioactive C_{14} has a half-life time $T_{1/2}$ of 5730 years. At $t = 0$, $N(0) = N_0 = 14$ disintegrations per minute [dpm] per gram of natural carbon (Fig. 9.6).

Suppose an old artifact is found and its $C_{14} = 4$ [dpm]. The original date of the artifact can be approximated between 9500 and 10,000 years using Eq. (9.2). The same formula is used to assess the impact of nuclear waste.

1.3 Oscillation

The third basic mode of behavior is called *oscillation*. This behavior is caused by *any delay* in a negative or balancing feedback loop. It can be a delay in measuring or reporting the discrepancy between the state of the system and the desired state (or goal) of the system, a delay in taking action after the discrepancy is determined, a delay between the corrective action based on the discrepancy measurement and its effect on the state of the system, or a combination of some of the aforementioned delays (Fig. 9.7).

A typical business-related example is the evolution of the unemployment rate and GDP growth. Unemployment is a lagging indicator. Employers start hiring new full-time workers once they are certain that the economy has been recovered over a period of time. Therefore, the unemployment rate may not decline until months after the economy starts to recover, as depicted in Fig. 9.8.

2 Interaction of the Basic Modes

In the previous section, we discussed the three basic modes of behavior and their underlying feedback structure: exponential growth, goal seeking, and oscillation. In this section, we will discuss some common modes of behavior that arise when the basic modes of behavior interact with each other. They include *S-shaped growth*, *S-shaped growth with overshoot*, S*-shaped growth with overshoot and collapse,* and *tragedy of the commons*. These modes of archetype behavior are dominantly present in the modeling of sustainable development and sustainable business process management.

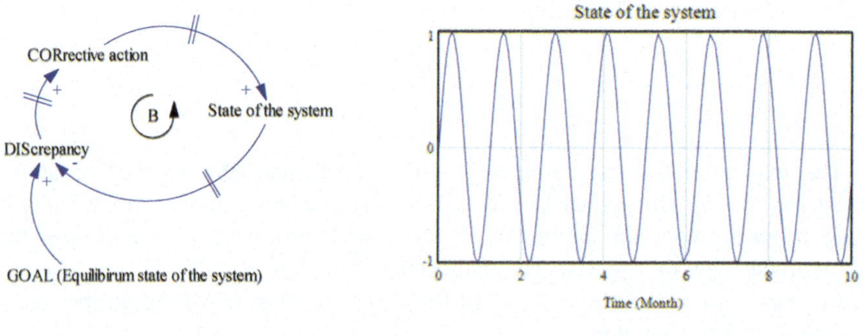

Fig. 9.7 State and structure of an oscillating system – CLD and illustrative example

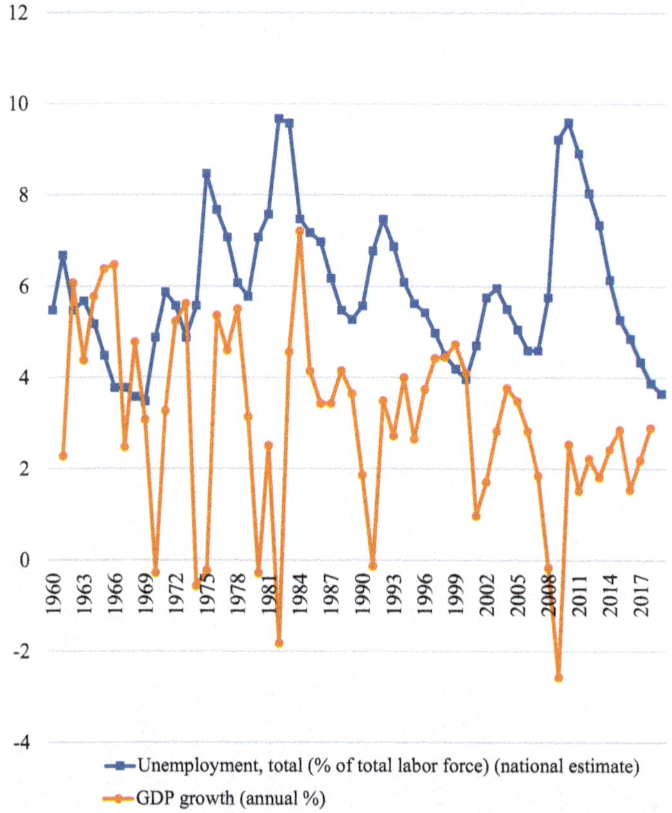

Fig. 9.8 Unemployment rate and annual GDP growth in the U.S. (Source: https://data.worldbank.org/indicator/)

2.1 S-Shaped Growth

We discussed exponential growth in Sect. 1.1. In real life, there are multiple examples of growth starting exponentially, but where one or more balancing or negative feedback loops eventually become dominant when the limit to growth is approaching.

This commonly observed mode of behavior is called *S-shaped growth*. Figure 9.9 demonstrates the behavior of S-shaped growth. Exponential growth is initially dominant, but after some time goal seeking behavior becomes dominant when the system faces the limit of growth, also known as its *carrying capacity*. The carrying capacity is the number of elements (people, cars, cell phones, etc.) that a particular system can support; it is determined by the resource availability and the resource requirements of the population.

Any real exponential growth behavior is limited by the exhaustion of resources in the environment needed for this growth. When the environment nears the carrying

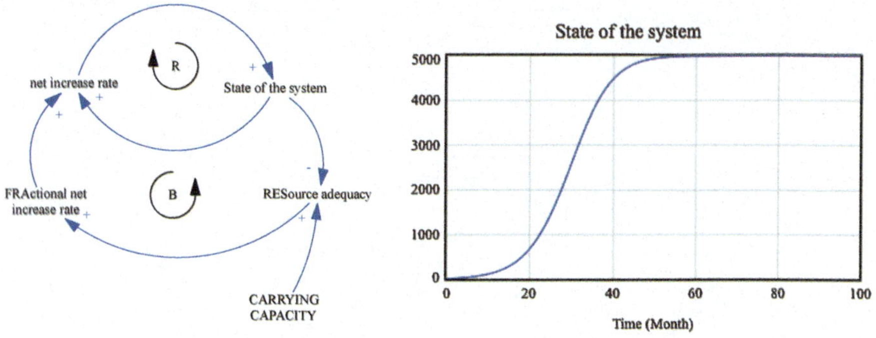

Fig. 9.9 State and structure of an S-shaped Growth system; example with carrying capacity of 5000 units

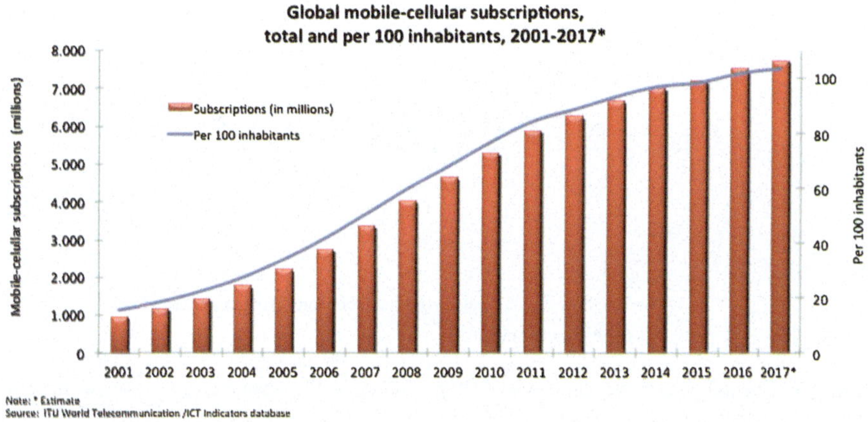

Fig. 9.10 Global number of yearly mobile phone subscriptions showing S-shaped growth behavior over period 2001–2017

capacity, the *adequacy* of the resources required to grow diminishes. As a result, the *fractional net increase rate* declines and finally becomes zero as soon as the carrying capacity is reached.

Please note that two conditions must be met for a system to show pure S-shaped growth: (i) the negative loops may not show any significant time delays and (ii) the carrying capacity must be fixed. As we will discuss in the next two sections, these two conditions are not always met in many real-life systems in the domain of sustainable development and sustainable supply chain management.

S-growth behavior is often observed during the transition phase of major technologies. Figure 9.10 shows the evolution in ownership of the number of global mobile cellular phones. The same S-shaped growth behavior was noticed in the twentieth century during the first 20–30 years after the introduction of electrical power in households, refrigerators, color TVs, air conditioning, computers, and many other technological applications used in daily life.

The S-shaped growth relationship is sometimes modeled with the so-called *logistic growth* model. We demonstrate the mathematical relationship of S-shaped growth using the example of constrained population growth (see Sect. 1.2).

Any population with plenty of food, space to grow, and no threat from predators tends to grow at a rate that is proportional to the population. This is exponential growth. If the reproduction rate r (i.e., the annual birth rate) can be considered constant, this growth rate is represented by the formula of exponential growth

$$\frac{dP}{dt} = r \cdot P \tag{9.5}$$

where

P is the population as a function of time t,
r is the (constant) birth rate expressed in number of people per year

Integrating both sides of Eq. (9.5) results in Eq. (9.6) expressing the number of people in the population at time t, given an initial population of P_0 people at $t = 0$ and a constant birth rate r:

$$P(t) = P_0 \cdot e^{r \cdot t} \tag{9.6}$$

In real life, population growth stops because of limited resources. This is the carrying capacity of that population. In Fig. 9.11, the carrying capacity is K people. As soon as the number of people $P(t)$ in the population approaches the carrying

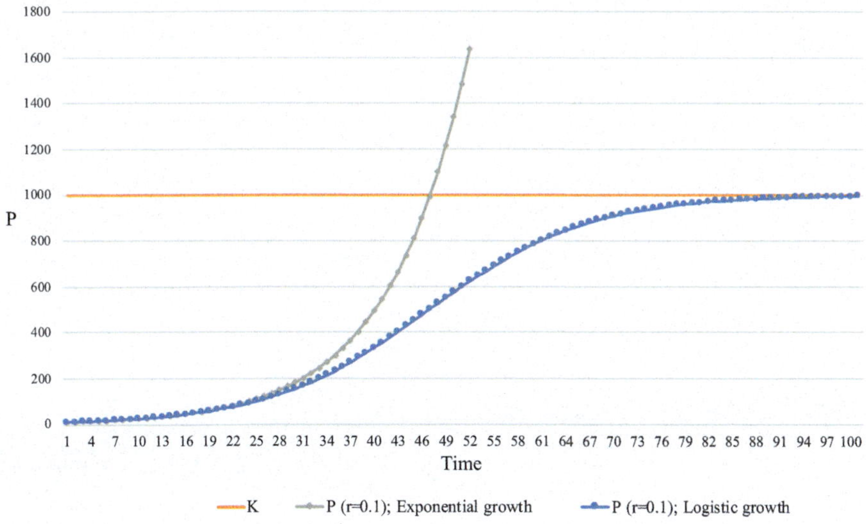

Fig. 9.11 How exponential growth changes towards S-shaped growth ($K = 1000$, $P_0 = 10$, $r = 0.1$)

capacity K, the growth rate drops to zero. The green curve demonstrates unlimited exponential growth of the population.

We may account for the growth rate declining to 0 by including a factor of $(1 - P(t)/K)$ in Eq. (9.5). When the population reaches the carrying capacity, i.e., when $P(t) = K$, this factor becomes 0. In the beginning, when the number of people in the population $P(t)$ is still small compared to the carrying capacity K, $P(t)/K$ can be considered as 0 representing the exponential growth given by (9.5). The resulting model (9.7) is called the *logistic growth model* or the *Verhulst model* named after the Belgian mathematician Pierre Verhulst who studied this kind of behavior in the nineteenth century.

$$\frac{dP}{dt} = r \cdot P(t) \cdot \left(1 - \frac{P(t)}{K}\right) \tag{9.7}$$

With P_0 being the initial population, the solution of Eq. (9.7) is:

$$P(t) = \frac{K \cdot P_0 \cdot e^{r \cdot t}}{K + P_0 \cdot (e^{r \cdot t} - 1)} \tag{9.8}$$

Using the data of the first five U.S. censuses, Verhulst predicted in 1840 how the U.S. population would evolve over the next century (from 1840 until 1940). The predictive ability of this formula was accurate until 1940, as shown in Fig. 9.12. After 1940, a deviation from the original carrying capacity can be noticed (Lipkin and Smith 2004). This example shows how significant effects, such as wars, can change the initial assumptions of a simple model.

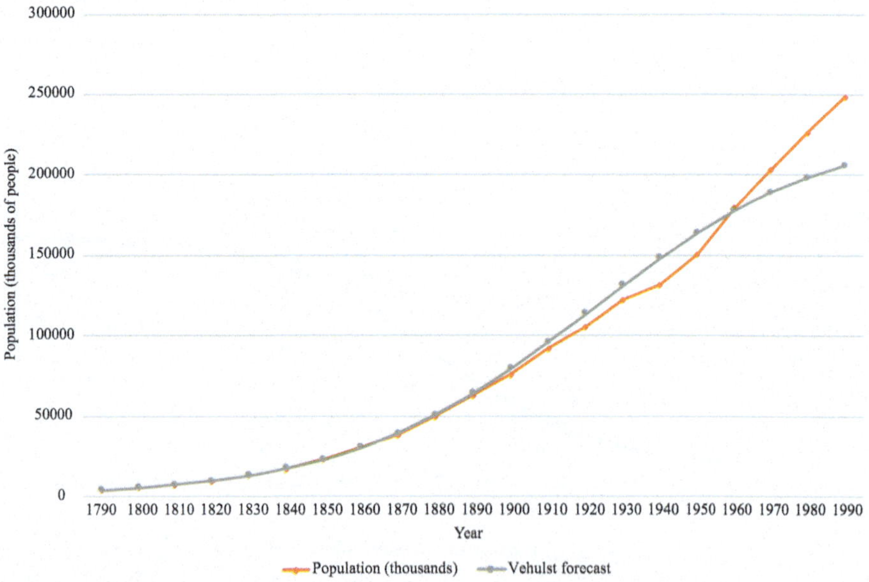

Fig. 9.12 Forecasted and historical evolution of U.S. population from 1780 to 2000. (Data from Lipkin and Smith 2004)

2.2 S-Shaped Growth with Overshoot

In the previous section, we discussed the two main prerequisites for S-shaped growth: no significant delays in the negative feedback loop and a fixed carrying capacity. In reality, delays may occur in the negative feedback loop of systems showing S-shaped growth. As a result, the system will overshoot and oscillate around the carrying capacity.

Delays may occur in the detection or the perception of the overshoot of the system. Another delay may occur in the adjustment of the fractional net increase rate. Both sources of delays are depicted in the CLD in Fig. 9.13. The example on the right side of Fig. 9.13 depicts the same system as pure S-growth of Fig. 9.9, taking a significant delay in the balancing feedback loop into account.

2.3 Overshoot and Collapse

Sometimes exceeding the carrying capacity can lead to an erosion of the carrying capacity. If this happens, the system collapses, as depicted in Fig. 9.14. After such a

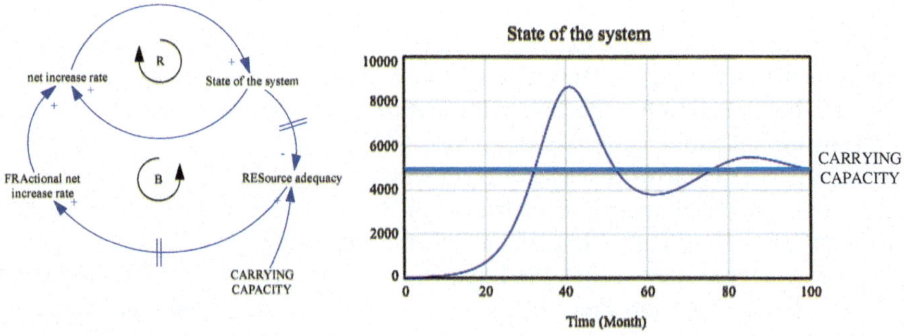

Fig. 9.13 State and structure of an S-shaped growth system with overshoot; example with carrying capacity of 5000 units

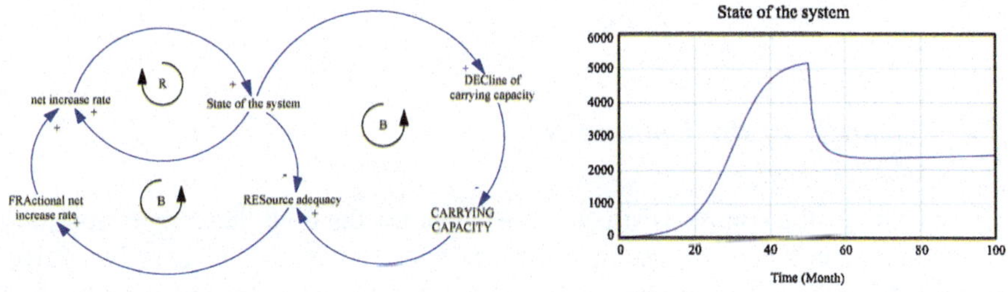

Fig. 9.14 State and structure of an S-shaped growth system with overshoot and collapse

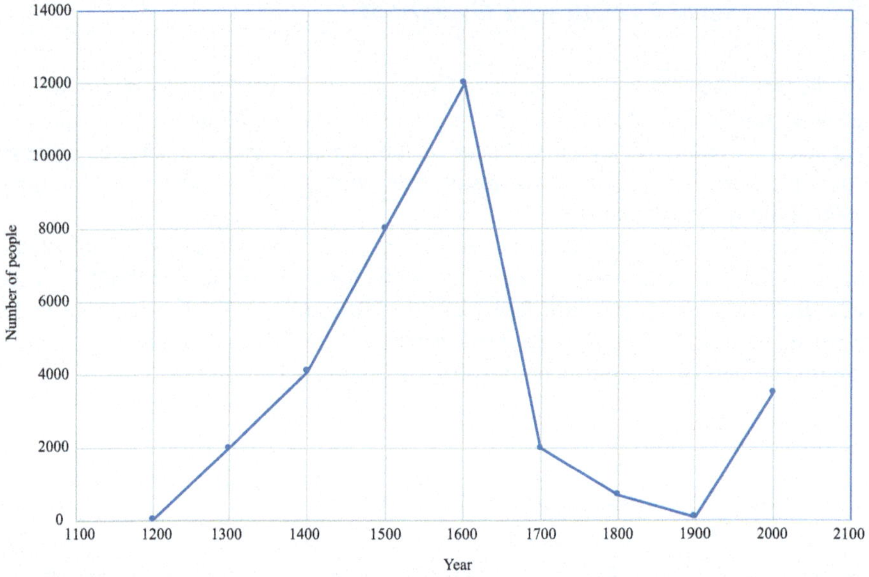

Fig. 9.15 Rise and Fall of the population of Easter Island: estimation of the population size over time

collapse, the new carrying capacity is smaller than the original one; this changes the future behavior of the system. The CLD example is the same one depicted in Fig. 9.9, but the erosion or decline of the carrying capacity starts when the carrying capacity of 5000 units is surpassed.

A famous example of S-shaped growth with overshoot and decline is the evolution of the population on Easter Island, an isolated and remote island located 3759 km east of Santiago, Chile, in the Atlantic Ocean. Recent C_{14} dating suggests that the first inhabitants, called the Rapa Nui, arrived around 1200 AD (Cole and Flenley 2008). The population grew fast, and by the late 1500s, an estimated 10,000–12,000 people lived on the island. This resulted in a surpassing of the island's carrying capacity, which was accelerated by soil erosion and the disappearance of palm trees. The island's ecology collapsed, and this was followed by civil war. Finally, the population diminished to a historic low of 111 inhabitants by 1877 (Fig. 9.15).

2.4 *Tragedy of the Commons*

The tragedy of the commons structural behavior results from the competition for a common resource between multiple competitors (Activity A and B in Fig. 9.16). Initially, the behavior of competitors *A* and *B* is not limited by the common resource limit. The individual reinforcing loops, in this initial phase, of competitors A and B

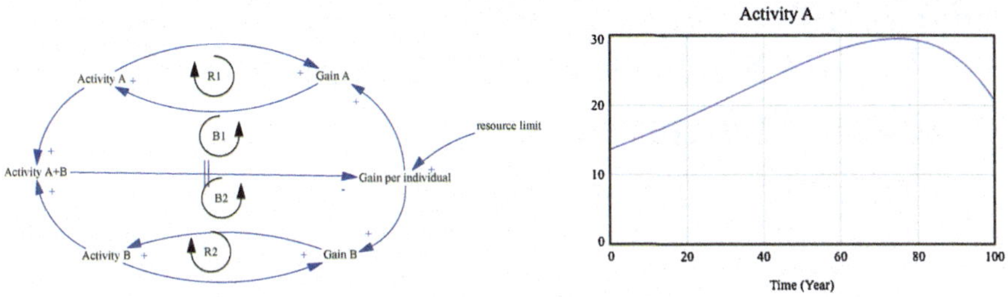

Fig. 9.16 State and structure of an archetype system showing tragedy of the commons behavior

result in the growth of Activities A and B. Due to the growing activities of A and B, the sum of both activities approaches the resource limit of the common resource. At that moment, the balancing structure of the common resource decreases Activities A and B. The gain per individual drops as the total activity approaches the resource limit. Figure 9.16 shows a typical behavior of how Activity A evolves over time. Once the resource limit is surpassed, Activity A decreases. A similar behavior applies to Activity B.

2.5 *Other Modes of Behavior*

There are many other archetypes of behavior. Some of them can be viewed on http://www.systems-thinking.org/arch/arch.htm

Chapter 10
Stocks and Flows and the Dynamics of Simple Structures

In the previous chapter, we discussed the structure and state of the dynamic behavior of simple archetype systems. We used CLDs to formalize the discussion about their behavior. We will go one step further in this chapter. By introducing new concepts like *stocks*, *flows,* and *delays*, we will be able to build models that enable us to analyze the dynamic behavior of systems.

The principles of system dynamics are based on two major systems principles. The first is that stocks, flows, and delays determine system behavior. The second is bounded rationality. System dynamics does not pretend to address all the variables of a problem; it concentrates on the ones that are key to the problem and its context, i.e., the "environment" as defined by the analyst. System dynamics does not pretend to optimize; it helps the analyst understanding a problem (Tang and Vijay 2001).

1 Introduction

Stocks and *flows* are used because causal loop diagrams cannot capture the difference between these central concepts. The difference between stocks and flows can easily be described by looking at the process of saving money in a bank savings account. Imagine that you save some money in your bank account every month. The amount of money saved increases every month. As soon as money is debited from the account, the net amount in the account decreases. The current amount of money saved in your bank account is considered a *stock variable,* and the inflow and outflow of money in the account via saving or withdrawal are the *flow variables*, which change the state of the stock of money in the bank account.

Stocks accumulate or integrate their flows. The net flow into a stock is the rate of change of the stock. In general, the relation between stocks and flows is given by the following integral Eq. (10.1) and differential Eq. (10.2):

D. Inghels, *Introduction to Modeling Sustainable Development in Business Processes*, https://doi.org/10.1007/978-3-030-58422-1_10

$$Stock(t) = \int_{t_0}^{t} [inflow(s) - outflow(s)] \cdot ds + Stock(t_0) \tag{10.1}$$

$$\frac{d(Stock)}{dt} = \text{Net Change in Stock} = inflow(t) - outflow(t) \tag{10.2}$$

Where:

Stock (t): stock present at time t
Inflow (s): the inflow at any time s between the initial time t_0 and the current time t
Outflow (s): the outflow at any time s between the initial time t_0 and the current time t

Consider Fig. 10.1: a box represents a stock, and a valve or bowtie represents a flow. Clouds represent the boundaries of the model. The cloud on the inflow represents a *source* of unlimited capacity, and the cloud on the outflow represents a *sink* of unlimited capacity.

Stocks are critical in generating the dynamics of systems for the following reasons (Sterman 2000):

- **Stocks characterize the state of the system** and provide the basis for actions. Stocks are like a company's balance sheet; they depict the current state of a system and express how the system behaves at a particular moment in time. Based on that information, actions can be taken if necessary.
- **Stocks provide systems with inertia and memory.** Take, for example, the current GHG emissions in our atmosphere. This stock is the accumulation of past events that produce greenhouse gasses. The content of this stock can only change due to changes in inflow or outflow. The current state of these GHG emissions can be viewed as a memory. In general, stocks do not have to be tangible. Memories and beliefs (see the paradigms discussed in Chap. 7), as an example, characterize the intangible mental state. Examples of tangible stocks are people, money, and material. Intangible stocks include perceptions, expectations, etc.

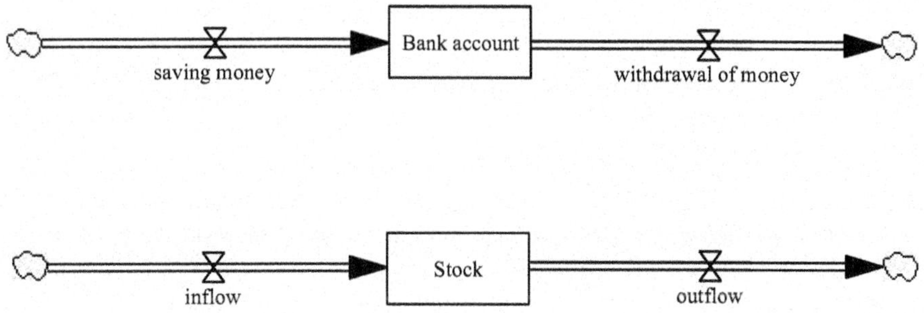

Fig. 10.1 Bank savings account metaphor for stocks and flows

- **Stocks are the source of delays.** A delay is a process whose output lags behind its input. The difference between the input and output accumulates in a stock of material.
- **Stocks** decouple rates of flow and **create disequilibrium dynamics.** Inflows and outflows can differ due to the stock they are connected to, which may be able to absorb the difference between outflow and inflow. In equilibrium, the inflow of a stock equals the outflow. However, this is often the exception. Most of the time, disequilibrium is the rule because different decision processes govern inflows and outflows.

Stocks and flows are present in many disciplines, but they are given different names. Examples of stocks in manufacturing are buffers or inventories of goods. The throughput rate of machines or a process are examples of flows in manufacturing. Stocks and flows can be defined in chemical, economic, accounting, environmental, societal, and other processes as well.

The units of measurement can help to distinguish stocks from flows. Stocks usually represent an amount, such as the total savings in your bank account expressed in [€], whereas the associated flows are measured in the unit but per time. The average saving amount per month is a flow expressed in [€/month], and the average amount of money withdrawn per month is an outflow expressed in the same dimension of [€/month].

2 Variables and Model Boundaries

A system consists of networks of stocks and flows linked by *information feedbacks* from stocks to the flows. The boundary of a system consists of *sources* and *sinks,* which are considered to have infinite capacity. In addition, to stocks and flows, systems can consist of *constants*, *auxiliary* and *exogenous variables.* Constants are variables that do not change or that change so slowly over the time of interest that they are considered to be steady. Exogenous variables are stocks outside the model boundary. Auxiliary variables are intermediate variables that are helpful for the clarity of the model. Ideally, each equation in a model should only represent one main item. It is advisable to split up a complex equation into separate parts using auxiliary variables so that the system is readable and understandable (Fig. 10.2).

Furthermore, some general conventions are not mandatory, but advisable to improve the legibility of an SD model:

Delays are depicted by a double bar on the causal link representing the delay

Stock variables are written with initial capital letters on each word such as Stock1 and Stock2

Flow variables are written in lower case such as flow1, flow2, and flow3

Constants are written using all capital letters, such as CONSTANT

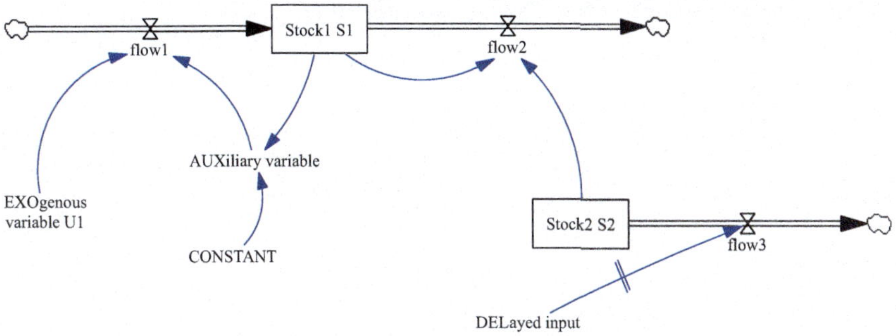

Fig. 10.2 System dynamics notation convention used in this book

Auxiliary and exogenous variables are written with the first three letters in capitals and the subsequent letters in lower case, such as AUXiliary variable and EXOgenous variable.

Causal links can also have a polarity assigned in a Stock and Flow diagram similar to a CLD.

In the next section, we will explore two simple *linear first order* feedback systems to discuss elementary systems connecting stock and flow. *Linear* relates to the fact that the corresponding rate equation, i.e., the net inflow to a stock, is a linear combination of state variables (S_i) and exogenous inputs (U_i) (Sterman 2000):

$$\frac{dS}{dt} = \text{Net Inflow} = \mathrm{a}_1 \cdot \mathrm{S}_1 + \mathrm{a}_2 \cdot \mathrm{S}_2 + \cdots + \mathrm{a}_\mathrm{n} \cdot \mathrm{S}_\mathrm{n} + \mathrm{b}_1 \cdot \mathrm{U}_1 + \mathrm{b}_2 \cdot \mathrm{U}_2 + \cdots + \mathrm{b}_\mathrm{n} \cdot \mathrm{U}_\mathrm{n} \tag{10.3}$$

where coefficients a_i and b_i are constants. The *order* of a system relates to the number of stocks, i.e., the number of state variables the system contains. In a first order linear system, only one stock S is present, and the net inflow rate is a linear relationship with this stock or state variable.

3 Dynamics of Exponential Growth

Let's look at a linear first order system that shows exponential growth behavior (see Chap. 9 for the CLD). Please recall that linear first order systems showing exponential behavior have a positive feedback loop and one state variable or Stock S. The net inflow is directly proportional to the state of the system (Fig. 10.3).

The proportionality factor g represents the fractional growth rate of the stock and has the dimension [1/time]:

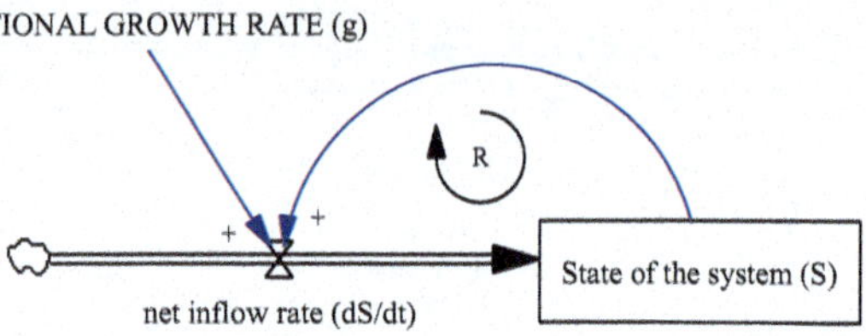

Fig. 10.3 Stock and flow diagram of a linear first order system showing exponential behavior

$$\frac{dS}{dt} = g \cdot S \tag{10.4}$$

This differential equation is solved by reorganizing Eq. (10.4):

$$\frac{dS}{S} = g \cdot dt \tag{10.5}$$

After integrating both sides, we obtain the expression representing the State S of this exponential growth system as follows:

$$\int \frac{dS}{S} = \int g \cdot dt \tag{10.6}$$

Solving these integrals delivers the following expression in which C represents a constant:

$$\ln(S) = g \cdot t + C \tag{10.7}$$

This is expression is equal to:

$$S(t) = e^{g \cdot t} \cdot e^{C} \tag{10.8}$$

At time $t = 0$: $S(t) = e^{0}.e^{c} = 1 \cdot S(0) = S(0)$.

Therefore, in general the state of a first order linear system with exponential growth behavior is characterized by the following expression:

$$S(t) = S(0) \cdot e^{g \cdot t} \tag{10.9}$$

Exponential growth has the property that during a fixed period of time, called the *doubling time*, t_d, the state of the system doubles. Using Eq. (10.9): $2 \cdot S(0) = S(0) \cdot e^{g \cdot t_d}$, the doubling time can be expressed as:

$$t_d = \frac{\ln(2)}{g} \approx \frac{0.7}{g} \tag{10.10}$$

Take as an example the average real GDP growth rate in the US over the past 100 years. This rate has been 3.4%/year, which relates to a doubling time $t_d = \frac{0.7}{0.034} \approx 20$ [year]. In other words, over the past 100 years, the GDP in the US has doubled on average every 20 years.

It is very hard for us to become accustomed to exponential growth because we have a linear thinking mindset. The following experiment illustrates this: take a sheet of paper. It usually has a thickness of 0.1 mm. Fold it double. The thickness is now 0.2 mm. Fold it again. The thickness will then be 0.4 mm. After 17 folds, the thickness has already grown to $2^{17} \cdot 0.1$ mm = 13 m. At 100 folds, it has the radius of the universe (view: https://thesystemsthinker.com/paper-fold-an-exercise-in-exponential-growth/ to see more). Another example that illustrates that we are not familiar with exponential growth is the sudden rise in 2020 of positively confirmed cases of COVID-19 in many countries. The coronavirus spread exponentially in many countries, and that surprised many people.

We take the example of fixed compound interest to show how exponential behavior is modeled in Vensim®. We use Vensim® (https://vensim.com) in this book to simulate system dynamics models. The stock variable in Fig. 10.4 is called "Deposit Account". This is a stock variable since an amount of money is accumulated over time. The inflow of money to the deposit account is represented by the inflow variable "annual net increase rate", which adds a value equal to the product of the interest rate to the current state of the value in the deposit account every year.

The Vensim® program code looks like this:

(01) annual net increase rate = INTEREST RATE*Deposit Account
Units: Euro/year

(02) Deposit Account = INTEG (annual net increase rate, 100)
Units: Euro

(03) FINAL TIME = 35
Units: year
The final time for the simulation.

Fig. 10.4 Stock and flow diagram of a deposit account with a fixed annual interest rate

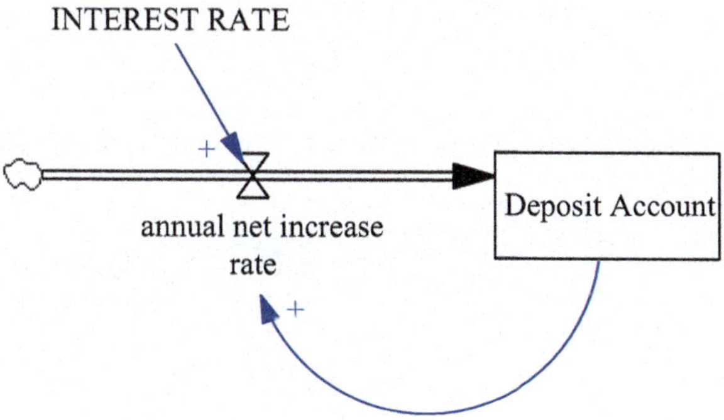

(04) INITIAL TIME = 0
Units: year
The initial time for the simulation.

(05) INTEREST RATE = 0.02
Units: 1/year

(06) SAVEPER = TIME STEP
Units: year [0,?]
The frequency with which output is stored.

(07) TIME STEP = 0.125
Units: year [0,?]
The time step for the simulation.

The evolution of the stock variable "Deposit Account" is depicted in Fig. 10.5. This is exactly the same as Fig. 9.2. Note that after 35 years, the amount of the deposit account has doubled. The proportionality factor in this example is the annual constant interest rate of 2%. Using Eq. (10.10), the doubling time is $t_d \approx \frac{0.7}{0.02} \approx 35$ years.

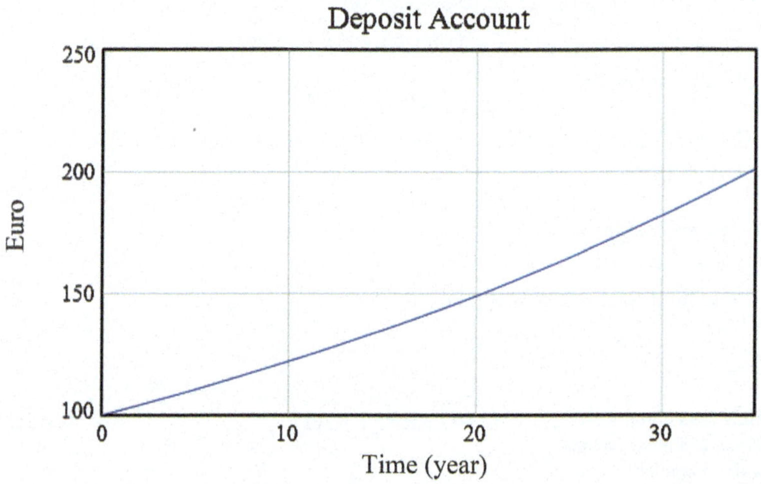

Fig. 10.5 State of the stock variable "Deposit Account" over time

4 Dynamics of Goal-Seeking Behavior

In the previous section, we discussed first order linear *positive* feedback systems that generate exponential growth behavior. In this section, we will discuss first order linear *negative* or balancing feedback systems that generate goal-seeking behavior.

The general stock and flow diagram of a linear first order system showing goal-seeking behavior is depicted in Fig. 10.6. The current state of the system is S, and the *desired state* is S^*. The time constant for the loop is called the *adjustment time AT*.

The net rate of change of the single stock in this system is linear:

$$\frac{dS}{dt} = \text{Net Inflow} = \frac{(S^* - S)}{AT} \tag{10.11}$$

Where AT is the adjustment time expressed in [time].

The deduction of the stock behavior over time is similar to the behavior of exponential growth. Finally, the result is:

$$S(t) = S^* - (S^* - S(0)) \cdot e^{\frac{-t}{AT}} \tag{10.12}$$

Linear first order systems showing goal-seeking behavior have a *desired state* S^* that will be met by the state of system S after a theoretical indefinite period of time. This is the equilibrium state of the system. The speed at which system state S will be equal to the desired state S^* is given by the *adjustment time AT* or time constant of the system.

Figure 10.7 gives more insight into Eq. (10.12) for $S^* = 1$, $S(0) = 0$ and $AT = 10$ s. First order linear systems showing goal-seeking behavior will eventually meet the desired state by systematically closing the gap between the current state S and the desired state S^*. This gap is closed exponentially. It goes fast in the beginning, but it slows down when the actual state S reaches the desired state S^*. Theoretically, the gap is never closed, but practically one may assume that the gap is closed after five times AT. At that point in time, the remaining gap is still 0.67% of the initial gap at time $t = 0$ (see Table 10.1).

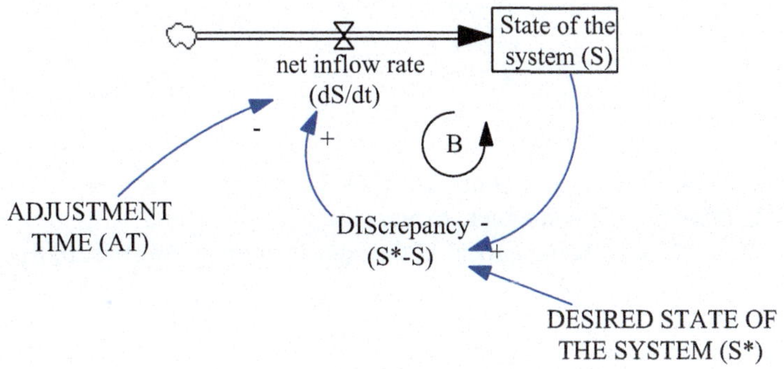

Fig. 10.6 Stock and flow diagram of a linear first order system, showing goal seeking behavior

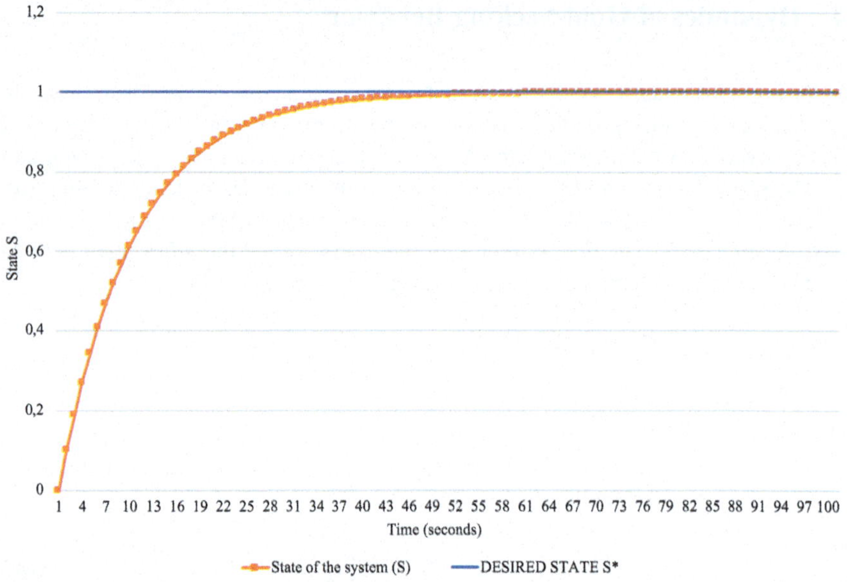

Fig. 10.7 Dynamic behavior over time of a linear first order system showing goal-seeking behavior over time. ($S^* = 1$, $S(0) = 0$, $AT = 10$ s)

Table 10.1 Remaining gap over time for a system with goal-seeking behavior ($S^* = 1$, $S(0) = 0$, $AT = 10$ s)

Time *t* [seconds]	Time [#AT]	Fraction of initial gap corrected	Fraction of initial gap remaining
0	0	0.00%	100.00%
10	1	63.21%	36.79%
20	2	86.47%	13.53%
30	3	95.02%	4.98%
40	4	98.17%	1.83%
50	5	99.33%	0.67%
60	6	99.75%	0.25%
70	7	99.91%	0.09%
80	8	99.97%	0.03%
90	9	99.99%	0.01%
100	10	100.00%	0.00%

A special form of goal-seeking behavior is called *exponential decay,* where $S^* = 0$. In this case, it is common to express the adjustment time AT as the *fractional decay* rate d, which is the reciprocal of the adjustment time AT (Fig. 10.8).

$$d = \frac{1}{AT} \; [1/\text{time}] \tag{10.13}$$

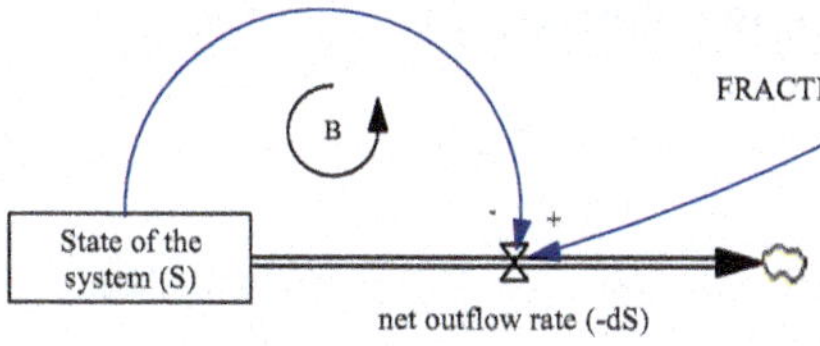

Fig. 10.8 Stock and flow diagram of a first order system with exponential decay behavior

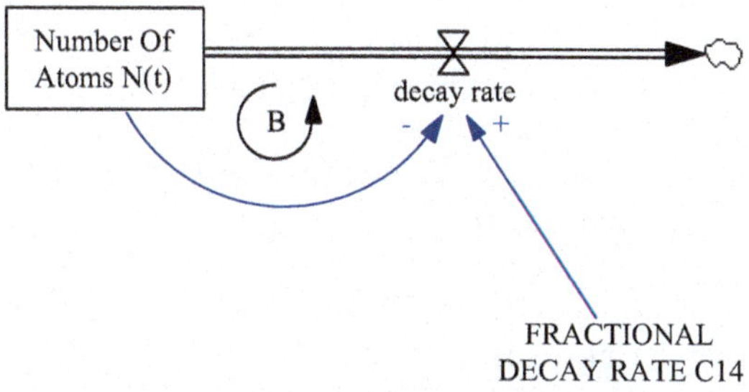

Fig. 10.9 Stock and flow diagram of the exponential decay behavior of C_{14}

Equation (10.12) can be rewritten for exponential decay as follows using (10.13) and knowing that $S^* = 0$:

$$S(t) = S(0) \cdot e^{-dt} \tag{10.14}$$

Exponential decay has a unique property; over a fixed period of time, called the *half-life time* t_h, the state of the system is cut by 50%. Using Eq. (10.14): $0.5 = e^{-\frac{t_h}{AT}}$, the half-time can be expressed as:

$$t_h = AT \cdot \ln(2) \approx 0.7 \cdot AT \approx \frac{0.7}{d} \tag{10.15}$$

In other words, after 70% of the period equal to AT the state S of the system is half of the initial state, and after one time period equal to AT, the remaining gap is $e^{\frac{-AT}{AT}}$ or 37% of its initial value. This means that after a time period AT, 63% of the gap is closed compared to the previous time period.

Let's re-examine the C_{14} radioactivity example that we discussed in Chap. 9 and look at how to model this example of exponential decay in Vensim®. The decay rate of C_{14} is $d = 0.000121$ [1/year] so following (10.13), $AT = 8267$ [year]. The initial number of atoms for C_{14} at $t = 0$ is expressed by $N_0 = 14$ [dpm]. The number of atoms decreases following an exponential decay pattern depicted in the stock and flow diagram in Fig. 10.9.

The corresponding Vensim® program code looks like this:

(01) decay rate = FRACTIONAL DECAY RATE C14* "Number Of Atoms N(t)"
Units: dpm/year

(02) FINAL TIME = 15000
Units: year
The final time for the simulation.

(03) FRACTIONAL DECAY RATE C14 = 1/8267
Units: 1/year

(04) INITIAL TIME = 0
Units: year
The initial time for the simulation.

(05) "Number Of Atoms N(t)" = INTEG (-decay rate, 14)
Units: dpm

(06) SAVEPER = TIME STEP
Units: year [0,?]
The frequency with which output is stored.

(07) TIME STEP = 0.125
Units: year [0,?]
The time step for the simulation.

Note that the exponential decay curve depicted in Fig. 10.10 is exactly the same as the one depicted in Fig. 9.6.

A more general example of a system with goal-seeking behavior is a thermostat controlling the temperature in a room. Suppose the initial room temperature is 15 °C and the adjustment time of the system is 0.5 h. The stock and flow diagram of this system is shown in Fig. 10.11. The controller steers the heating device proportional to the temperature gap to overcome the gap between the desired room temperature, which is 20 °C, and the current room temperature of 15 °C.

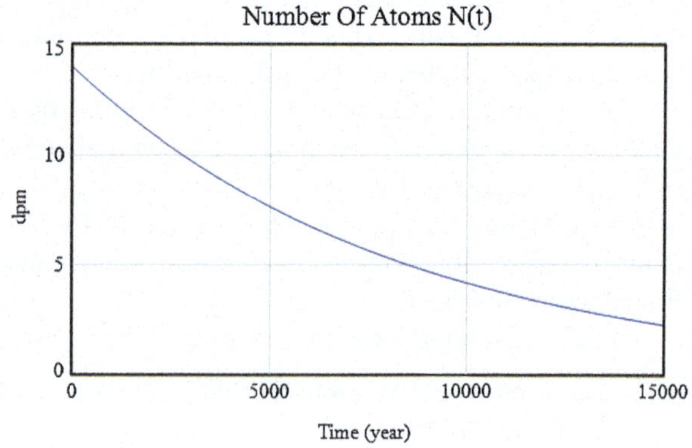

Fig. 10.10 Exponential decay evolution of C_{14} over time

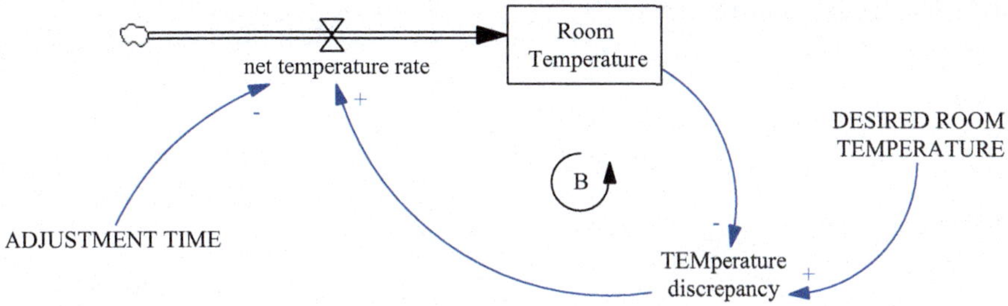

Fig. 10.11 Stock and flow diagram of controlling the room temperature

The corresponding Vensim® program code looks like this:

(01) ADJUSTMENT TIME = 0.5
Units: Hour

(02) DESIRED ROOM TEMPERATURE = 20
Units: Degree C

(03) FINAL TIME = 10
Units: Hour
The final time for the simulation.

(04) INITIAL TIME = 0
Units: Hour
The initial time for the simulation.

(05) net temperature rate = TEMperature discrepancy/ADJUSTMENT TIME
Units: Degree C/Hour

(06) Room Temperature = INTEG (net temperature rate, 15)
Units: Degree C

(07) SAVEPER = TIME STEP
Units: Hour [0,?]
The frequency with which output is stored.

(08) TEMperature discrepancy = DESIRED ROOM TEMPERATURE-Room Temperature
Units: Degree C

(09) TIME STEP = 0.125
Units: Hour [0,?]
The time step for the simulation.

The dynamic behavior or evolution of the room temperature is depicted in Fig. 10.12.

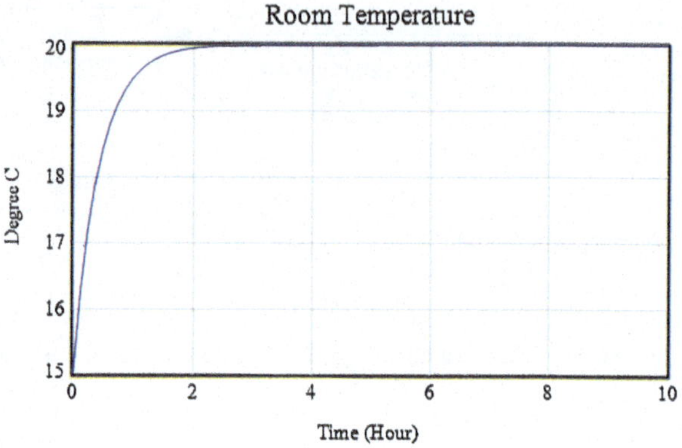

Fig. 10.12 Evolution of the room temperature over time

5 Dynamics of S-Shaped Growth

In S-shaped growth, exponential growth is dominant at first, but when the gap with the carrying capacity becomes smaller, the system predominantly shows goal-seeking behavior before finally reaching the carrying capacity (see Chap. 9).

In pure S-shaped growth, the carrying capacity is fixed and can never be exceeded.

A typical example of S-shaped growth behavior can be observed in many countries; it is the accumulation of the national car stock over time. An initial increase in GDP drives the exponential sale of cars and the growth of the national car stock. After a while, this growth is limited by the country's ability to host all the cars. Figure 10.13 shows the stock and flow diagram of the increase of cars in a national car stock. The initial national car stock in this example is 200 cars/1000 people. The country has an infrastructure that can accommodate 700 cars/1000 people. This limit is called the carrying capacity of the national car stock (see Inghels et al. 2016a for more details).

Initially, when there is still a considerable gap between the current state of the national car stock and the desired state of 700 cars/1000 people, the reinforcing loop (depicted with the letter *R*) in Fig. 10.13 that encompasses the loop "National Car Stock Per 1000 People – net car stock increase rate" is dominant. Therefore, the state of the national car stock shows exponential growth. Once the gap between the current and desired state becomes smaller, the balancing loop (depicted with the letter *B*) encompassing the loop "National Car Stock Per 1000 People-RESource adequacy – FRActional net car stock increase rate – net car stock increase rate" becomes dominant and the state of the system shows goal-seeking behavior towards the desired state of 700 cars/1000 people (Fig. 10.14).

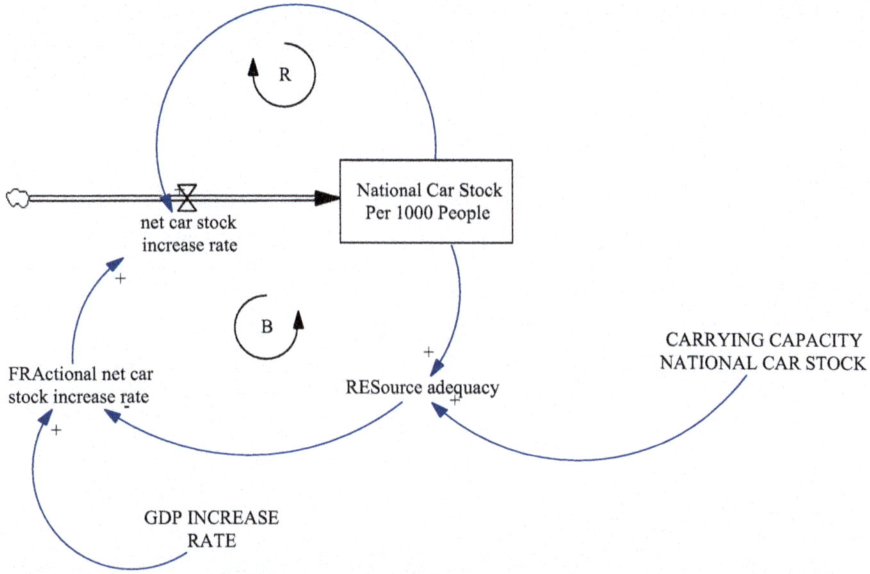

Fig. 10.13 Stock and flow diagram modeling the car increase in a national car stock based on Inghels et al. (2016a)

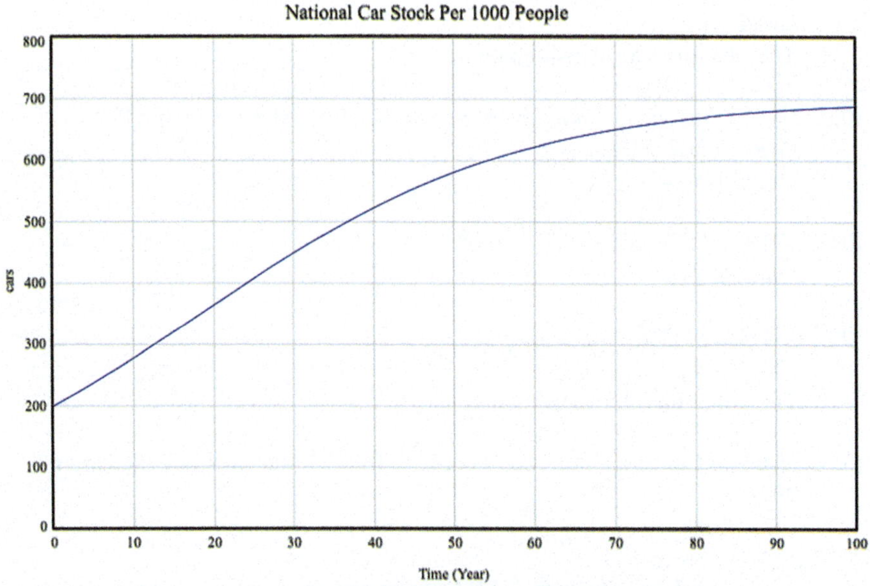

Fig. 10.14 National car stock evolution showing S-shaped growth behavior

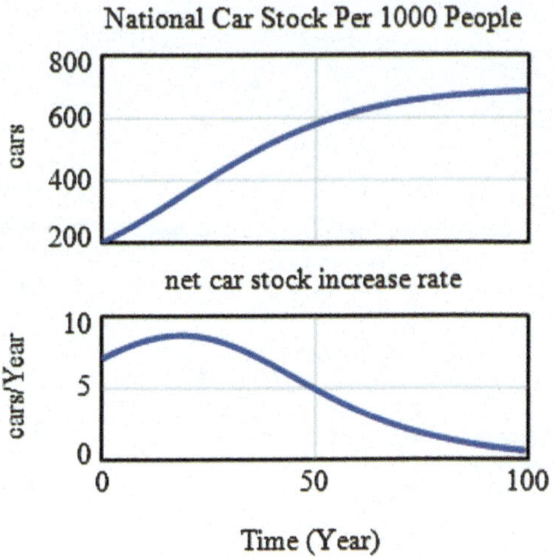

Fig. 10.15 Causal dependency "net car stock increase" – "National Car Stock Per 1000 People"

The Vensim® program code for this example looks like this:

(01) CARRYING CAPACITY NATIONAL CAR STOCK = 700
Units: cars

(02) FINAL TIME = 100
Units: Year
The final time for the simulation.

(03) FRActional net car stock increase rate = GDP INCREASE RATE*(1-RESource adequacy)
Units: 1/Year

(04) GDP INCREASE RATE = 0.05
Units: 1/Year

(05) INITIAL TIME = 0
Units: Year
The initial time for the simulation.

(06) National Car Stock Per 1000 People = INTEG (net car stock increase rate, 200)
Units: cars

(07) net car stock increase rate = FRActional net car stock increase rate*National Car Stock Per 1000 People
Units: cars/Year

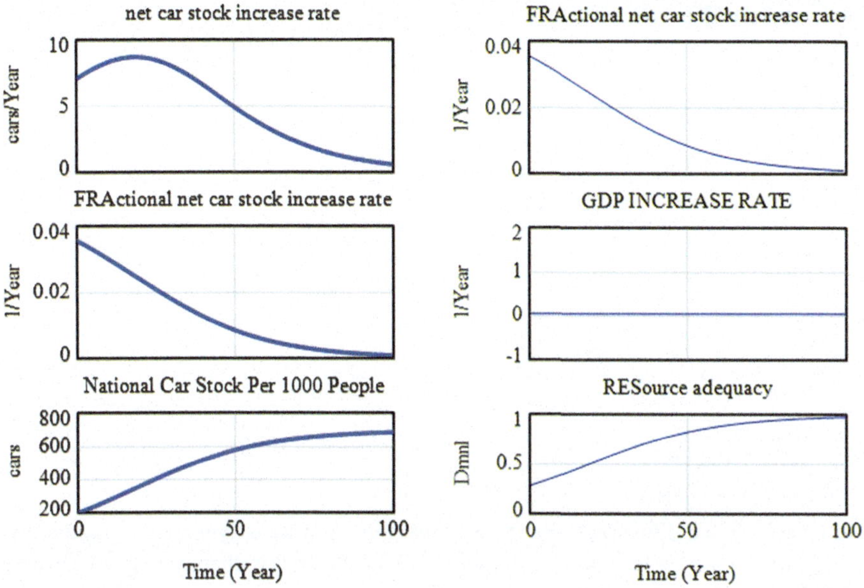

Fig. 10.16 Causal dependency "net car stock increase rate" with decreasing gap current-desired state

(08) RESource adequacy = National Car Stock Per 1000 People/CARRYING CAPACITY NATIONAL CAR STOCK
Units: Dmnl

Figure 10.15 shows how the inflow "net car stock increase rate" changes over time. It is the value that is integrated over time and added to the "National Car Stock Per 1000 People".

Figure 10.16 shows that the net car stock increase rate is determined by the fractional net car stock increase rate, which decreases because the gap with the carrying capacity, which is the desired state of this system, is getting smaller due to the increased number of cars in the national car stock.

Chapter 11
Delays

1 Introduction

In Chap. 10, we discussed three major elements of system dynamics to model real system behavior: stocks, flows, and feedback loops. In this chapter, we will discuss the fourth and last major element, *delay*. Whenever they become substantial, delays can create instability and oscillation in systems. Since a delay is a process whose output lags behind its input, at least one stock must be involved to decouple the input from the output of the stock. We recognize two types of delays: *material* and *information delays*. Within these two types of delays, we can recognize additional differences. First, we will discuss material delays, then information delays, before finally discussing the dynamics of two archetype system behaviors using delays: oscillating behavior and limits to growth behavior.

2 Pipeline (Material) Delay

The simplest form of a material delay is called a *pipeline delay*. In a pure pipeline delay, the outflow is exactly the same as the inflow but delayed with the *average delay time D*. In other words, for a given stock of Material in Transit, the relationship between inflow and outflow is given by:

$$\text{Outflow}(t) = \text{Inflow}(t - D) \tag{11.1}$$

The Material in Transit over the delay time D is the accumulated difference between outflow and inflow during the average delay time D added to the Material in Transit at the beginning of the pipeline delay at time t. Mathematically, this is expressed as follows:

D. Inghels, *Introduction to Modeling Sustainable Development in Business Processes*, https://doi.org/10.1007/978-3-030-58422-1_11

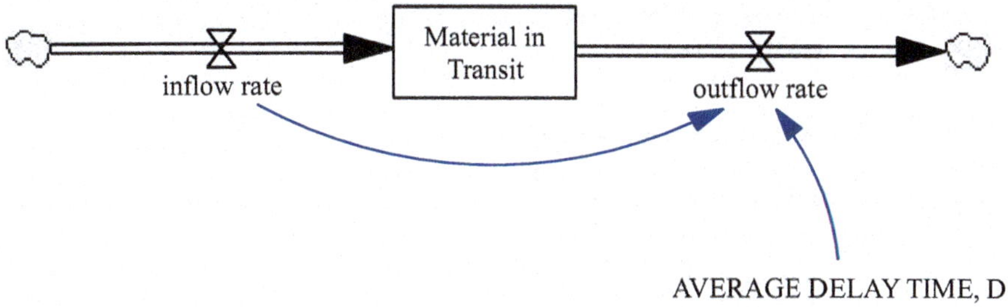

Fig. 11.1 Stock and flow diagram of pure (material) pipeline delay

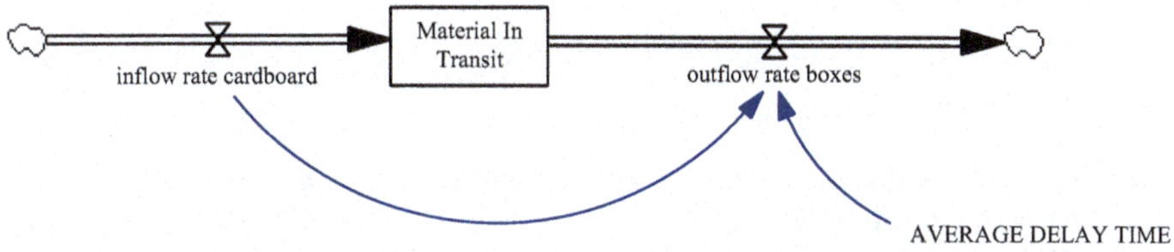

Fig. 11.2 Example pipeline delay

$$\text{Material in Transit} = \int_{t}^{t+D} (\text{Inflow}(t) - \text{Outflow}(t))dt + \text{Material in Transit }(t) \tag{11.2}$$

The stock and flow diagram of a typical pipeline delay is shown in Fig. 11.1.

In Vensim®, a pipeline delay is modeled using the function DELAY FIXED (*input*, *delay time, initial value*). The input, initial value, and output must all have the same units. The delay time must have the same units as TIME STEP. The input is the inflow rate of the system subjected to the pipeline delay with a delay time equal to the average delay time D from (11.1), and the initial value is the value of the outflow until $t = D$. The minimum delay time is the TIME STEP and shorter delay times will have the same effect as the delay time of TIME STEP.

An example will clarify how the DELAY FIXED function works. Take, as an example, a system of cardboard box production with a pipeline delay as depicted in Fig. 11.2. The inflow rate of cardboard sheet is 50 items/second from $t = 0$ onwards, and the outflow rate of cardboard boxes is initially 30 items/second. After an Average Delay Time = 10 s, the outflow rate equals the inflow rate and becomes 50 items/second. The stock "Material in Transit" is built up from 50–30 = 20 items at $t = 0$ to 220 items at $t = 10$ s. After this time (i.e., $t > 10$ s), the outflow rate equals the inflow rate, and the Material in Transit will remain constant (i.e., 220 Items) as long as no new changes occur.

The Vensim® program code for this example looks like this:

```
(01) AVERAGE DELAY TIME = 10
     Units: second
```

(02) FINAL TIME = 100
Units: second
The final time for the simulation.

(03) inflow rate cardboard = 50
Units: items/second

(04) INITIAL TIME = 0
Units: second
The initial time for the simulation.

(05) Material In Transit = INTEG (inflow rate cardboard-outflow rate boxes, 20)
Units: items

(06) outflow rate boxes = DELAY FIXED (inflow rate cardboard, AVERAGE DELAY TIME, 30)
Units: items/second

(07) SAVEPER = TIME STEP
Units: second [0,?]
The frequency with which output is stored.

(08) TIME STEP = 1
Units: second [0,?]
The time step for the simulation.

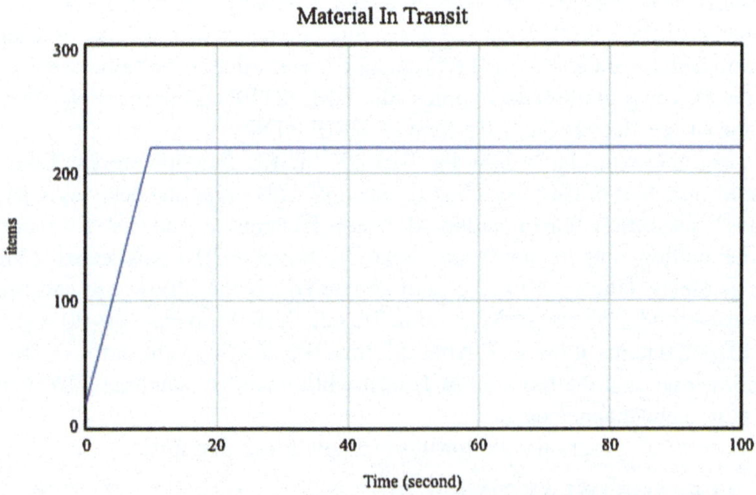

Fig. 11.3 Material in transit for the pipeline delay example of the production of cardboard boxes

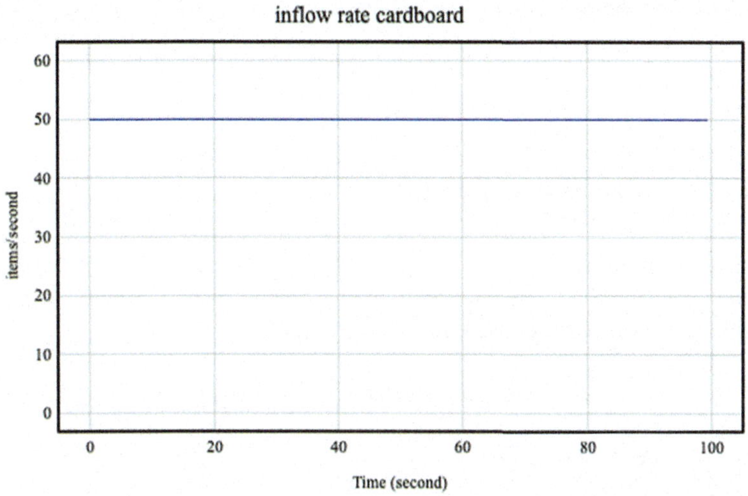

Fig. 11.4 Inflow rate pipeline delay example of the production of cardboard boxes

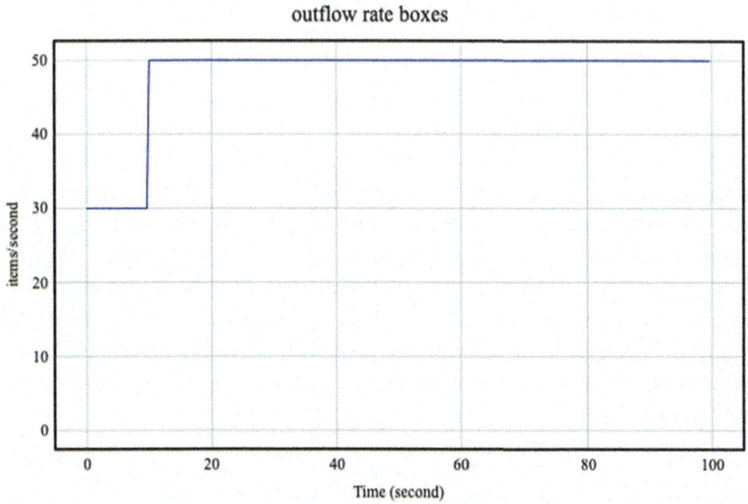

Fig. 11.5 Outflow rate for the pipeline delay example of the production of cardboard boxes with step change from 30 items/second to 50 items/second at time $t = 10$ s

An overview of how the flows and stocks in the example behave over time is given in Figs. 11.3, 11.4, and 11.5. Figure 11.3 shows how the Material in Transit is built up over time. In the period $t = 0$ s until $t = 10$ s, the outflow rate = 30 items/second and the inflow rate = 50 items/second. During 10 s, the Material in Transit is building up from 20 items to 20 + 10*(50–30) = 220 items.

The inflow rate over the period $t = 0$ s to $t = 100$ s is steady and equal to 50 items/second, as depicted in Fig. 11.4.

The outflow rate over the same period acquires two values: before $t = D$, the outflow rate is equal to the initial value of 30 items/second. At $t = D$, the outflow rate becomes equal to the inflow rate of 50 items/second.

3 First-Order Material Delay

The next type of delay is called a *first-order delay*. First-order refers to a system with one stock causing the delay. In a first-order material delay, the outflow is always proportional to the stock of Material in Transit (Sterman 2000):

$$\text{outflow} = \text{Material in Transit}/D \tag{11.3}$$

Where D = AVERAGE DELAY TIME [time]

$$\text{Material in Transit} = \int_{t=0}^{t} (\text{inflow rate}(t) - \text{outflow rate }(t)) \cdot dt + \text{Material in Transit }(O) \tag{11.4}$$

$$\text{Material in Transit}(0) = \text{inflow rate} \cdot D \tag{11.5}$$

Please note that the inflow is not present in Eq. (11.3). This means that the order of entry information regarding the individual items to the stock is not used to determine the outflow rate. The initial Material in Transit (0) Eq. (11.5) initializes the delay in equilibrium so that the initial outflow equals the initial inflow.

Since the inflow is not present in (11.3), the behavior of a first-order material delay, with the exception of the Material in Transit at $t = 0$ value (depicted with the grey arrow in Fig. 11.6), will be familiar to a system showing exponential decay as depicted in Fig. 10.8.

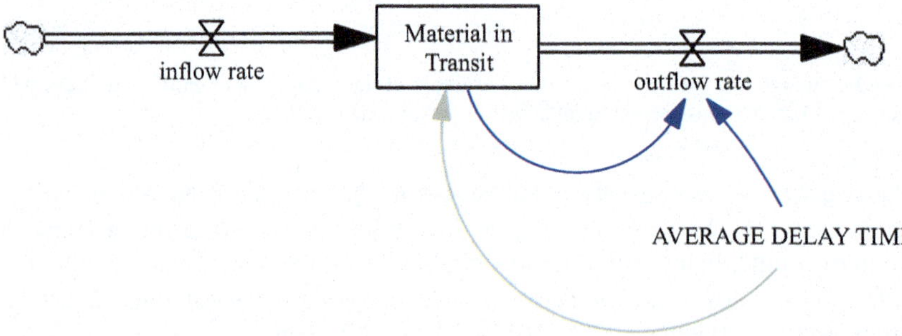

Fig. 11.6 Stock and flow representation of a system showing first-order material delay behavior

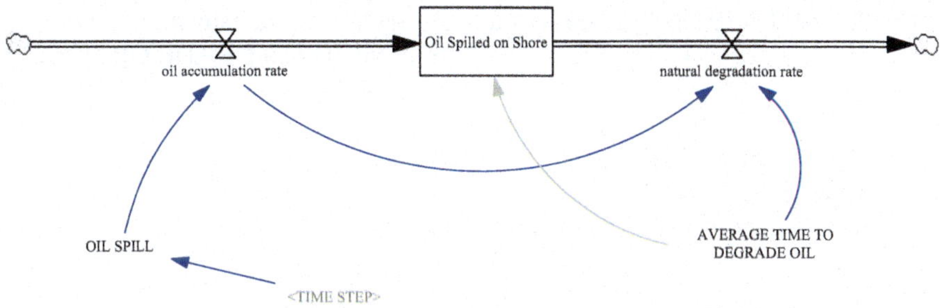

Fig. 11.7 Stock and flow representation of the natural degradation of oil spilled on shore based on MIT (1999)

The input for a first-order material delay system, the *inflow rate*, is the input to the delay. It can be a constant, a pulse, a sine wave, or any other function and is expressed in [Unit/time].

An example of a first-order material delay is the simple pollution degradation system (MIT 1999). The example describes how spilled oil, which is contaminating the shoreline, degrades over time when no action is taken. The spilled oil will naturally degrade and evaporate after a while. Figure 11.7 depicts the stock and flow model of this system.

The stock "Oil Spilled on Shore" represents the amount of spilled oil that reaches the shore. The oil accumulates at a rate equal to the "OIL SPILL". The oil will degrade naturally following a first-order material delay at a rate equal to the "Oil Spilled on Shore" divided by the "AVERAGE TIME TO DEGRADE OIL," which equals 6 years. After a certain length of time (equal to the "AVERAGE TIME TO DEGRADE OIL"), the oil levels on the shore should be at non-toxic levels.

The Vensim® program code of this model for a one-off oil spill at $t = 0$ is:

(01) AVERAGE TIME TO DEGRADE OIL = 6
Units: year

(02) FINAL TIME = 100
Units: year

(03) INITIAL TIME = 0
Units: year

(04) natural degradation rate = Oil Spilled on Shore/AVERAGE TIME TO DEGRADE OIL
Units: Gallon/year

(05) oil accumulation rate = OIL SPILL
Units: Gallon/year

(06) OIL SPILL = 5e+07*PULSE (0, TIME STEP)
Units: Gallon/year

(07) Oil Spilled on Shore = INTEG (oil accumulation rate-natural degradation rate, oil accumulation rate*AVERAGE TIME TO DEGRADE OIL)
Units: Gallon

(08) SAVEPER = TIME STEP
Units: year [0,?]
The frequency with which output is stored.

(09) TIME STEP = 0.0625
Units: year [0,?]
The time step for the simulation.

The oil spilled on shore in the model is considered to be a one-off event in the first year ($t = 0$) represented by a *Dirac impulse*. This is modeled in Vensim® with the function PULSE(_*start*_ , _*duration*_); *start* is the time the pulse starts and the *duration* of the pulse in time units is the second parameter to be filled in. In this example, the one-off oil spill takes place (i.e., starts) in year $t = 0$ and the duration is the smallest time period possible, i.e., one time step, in this case we choose 1/16 years (or 0.0625 years). The amplitude of the pulse, i.e., the pulse height, is a factor that should be multiplied with the function PULSE. In this case, we assume that 50,000,000 gallons of oil are spilled in a period of one time step pulse (see Fig. 11.9). To this end, the shadow variable <Time Step> is connected to the variable "OIL SPILL" in the stock and flow model of Fig. 11.7.

Figure 11.8 shows the state of the stock "Oil Spilled on Shore" over time. The initial value equals the 50,000,000 gallons of OIL SPILL at time $t = 0$ multiplied by the "AVERAGE TIME TO DEGRADE OIL" of 6 years. This equals the 300,000,000 gallons at t = 0 shown in Fig. 11.8. The state of the system over the

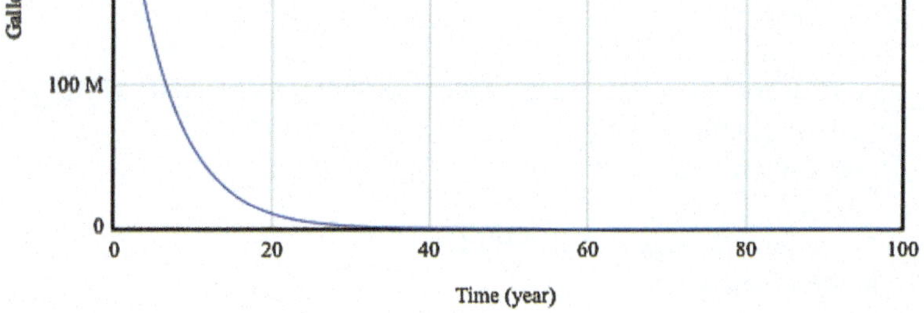

Fig. 11.8 State of "Oil Spilled on Shore" due to a Dirac impulse of 50,000,000 gallons at $t = 0$ year

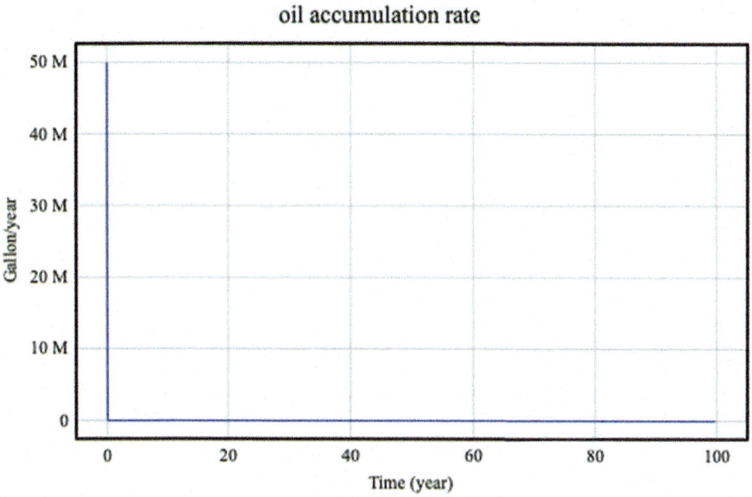

Fig. 11.9 Oil accumulation rate in the event of a one-off (Dirac impulse) oil spill of 50,000,000 gallons at $t = 0$ year

subsequent years shows an exponential decay behavior, as expressed in Fig. 11.8. Using Eq. (10.15), the half time is 4.2 [years] as shown in (11.6). After 4.2 years, the stock "Oil Spilled on Shore" is 150,000,000 [Gallons] (Fig. 11.9).

$$t_h = AT \cdot \ln(2) \approx 0.7 \cdot AT = 0.7 \cdot 6 = 4.2 \text{ [years]} \tag{11.6}$$

Following Eq. (11.3), the outflow is proportional to the stock of Material in Transit, in this case the stock "Oil Spilled on Shore". The proportionality factor is the reciprocal of the average delay time, D or AVERAGE TIME TO DEGRADE OIL in this case. Therefore, the behavior of the outflow "natural degradation rate" over time evolves exactly like the stock "Oil Spilled on Shore" but divided by a factor of 6 equal to the average time to degrade the oil as depicted in Fig. 11.10.

A first-order material delay can also be modeled using the Vensim® function DELAY1(*input rate*, *AVERAGE DELAY TIME*). This function returns an exponential delay of the input rate equivalent to the equations:

$$\text{DELAY1} = \text{Material in Transit}/\text{AVERAGE DELAY TIME}$$

$$\text{Material in Transit} = \text{INTEG (input rate} - \text{DELAY1, input rate} * \text{AVERAGE DELAY TIME)}$$

The input units should match the output units, and the units of the AVERAGE DELAY TIME must match those of the chosen TIME STEP.

For the pollution degradation problem, this becomes DELAY1 (*oil accumulation rate*, *AVERAGE TIME TO DEGRADE OIL*). If this function is modeled correctly, the state of the stock "Oil Spilled on Shore" and the state of the outflow "natural

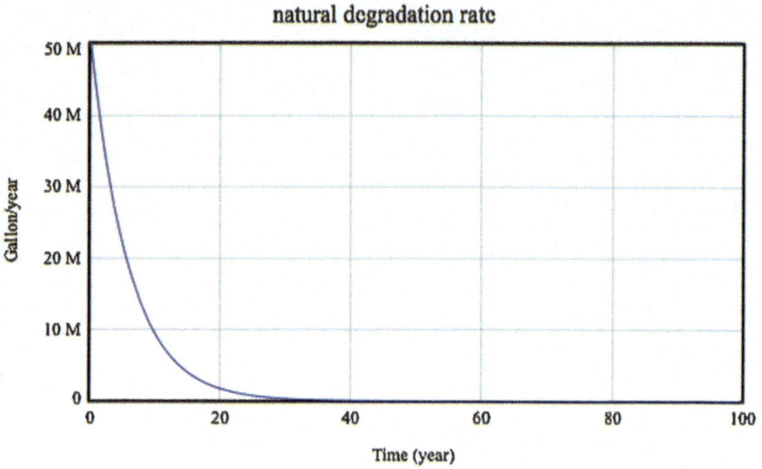

Fig. 11.10 First order material delay evolution of the natural degradation of a one-off oil spilled on shore event

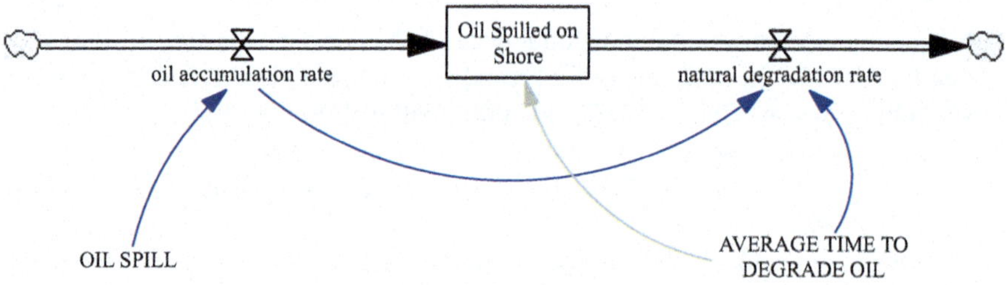

Fig. 11.11 Stock and flow diagram of the oil spill example modeled with Vensim® DELAY1 function

degradation rate" should be exactly the same over time, as depicted in Figs. 11.8 and 11.10. The adapted model using the function DELAY1 is shown in Fig. 11.11.

The Vensim® program code using DELAY1 looks now like this:

(01) AVERAGE TIME TO DEGRADE OIL = 6
Units: year

(02) FINAL TIME = 100
Units: year
The final time for the simulation.

(03) INITIAL TIME = 0
Units: year
The initial time for the simulation.

(04) natural degradation rate = DELAY1(oil accumulation rate, AVERAGE TIME TO DEGRADE OIL)
Units: Gallon/year

(05) oil accumulation rate = OIL SPILL
Units: Gallon/year

(06) OIL SPILL = 5e+07*PULSE (0, TIME STEP)
Units: Gallon

(07) Oil Spilled on Shore = INTEG (oil accumulation rate-natural degradation rate, oil accumulation rate*AVERAGE TIME TO DEGRADE OIL)
Units: Gallon

(08) SAVEPER = TIME STEP
Units: year [0,?]
The frequency with which output is stored.

(09) TIME STEP = 0.0625
Units: year [0,?]
The time step for the simulation.

The evolution over time of the inflow rate, outflow rate, and Material in Transit using the DELAY1 function for the oil spill example is depicted in Figs. 11.12,

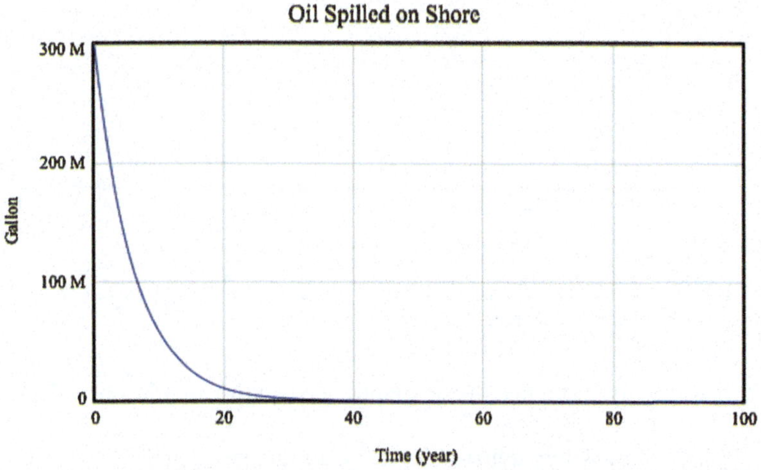

Fig. 11.12 State of the variable Oil Spilled on Shore in response to a Dirac inflow impulse in year $t = 0$ of 50,000,000 gallons modeled with the Vensim® function DELAY1

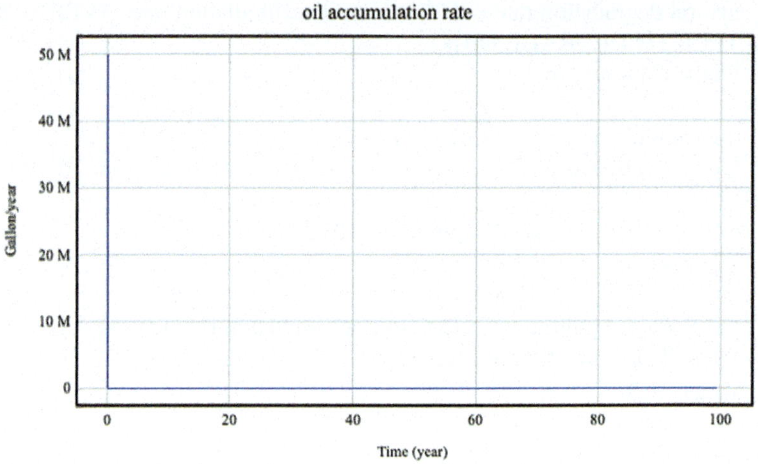

Fig. 11.13 Inflow rate modeled as a Dirac impulse of 50,000,000 gallons in year $t = 0$, modeled with the Vensim® function DELAY1

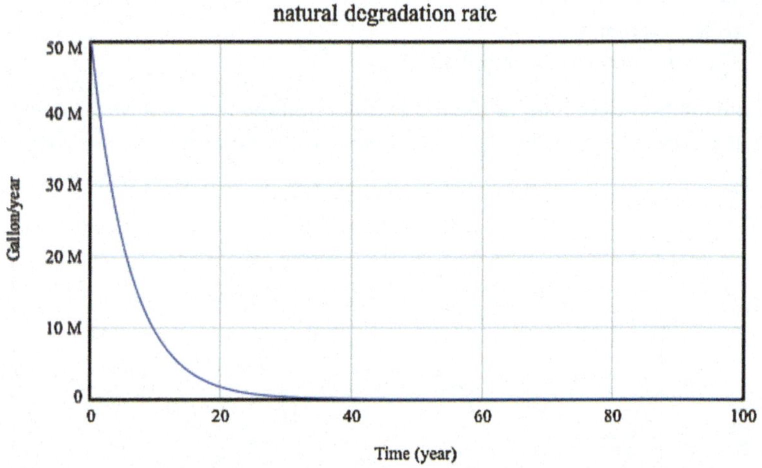

Fig. 11.14 Outflow rate in response to a Dirac impulse of 50,000,000 gallons in year $t = 0$, modeled with the Vensim® function DELAY1

11.13, and 11.14. Note that these graphs are similar to the ones depicted in Figs. 11.8, 11.9, and 11.10.

Both ways of modeling a first-order delay are compared in Table 11.1.

Table 11.1 shows the evolution of the outflow "natural degradation rate" for an initial pulse value (Dirac impulse) of 50,000,000 Gallon = 5 e+07 in year $t = 0$ that lasts one time step. After one average delay time period of 6 years (i.e., start of

Table 11.1 Response of the outflow "natural degradation rate" on a Dirac impulse of 5e+07 from $t = 0$ year until $t = 12$ year

Time t [Year]	Outflow "natural degradation rate" [Gallon/year] calculated as in Eq. 11.3–Eq. 11.5 using V	Fraction of "natural degradation rate" at time *t* compared with situation at $t = 0$	Outflow "natural degradation rate" [Gallon/year] modeled using the DELAY1 function in Vensim	Fraction "natural degradation rate" at time *t* using DELAY 1 function compared with situation at $t = 0$
0	5e+07	100%	5e+07	100%
1	4.27322e+07	85.46%	4.27322e+07	85.46%
2	3.57640e+07	71.53%	3.69053e+07	73.81%
3	3.02471e+07	60.49%	3.05655e+07	61.13%
4	2.55813e+07	51.46%	2.58503e+07	51.70%
5	2.16351e+07	43.27%	2.18629e+07	43.73%
6	1.82978e+07	36.60%	1.84904e+07	36.98%
7	1.54752e+07	30.95%	1.56381e+07	31.28%
8	1.30880e+07	26.18%	1.32258e+07	26.45%
9	1.10691e+07	22.14%	1.11856e+07	22.37%
10	9.36159e+06	18.72%	9.46013e+06	18.92%
11	7.91749e+06	15.83%	8.00083e+06	16.00%
12	6.76664e+06	13.53%	6.76664e+06	13.53%

$t = 6$ year), the outflow value "natural degradation rate" is 37% of the initial value. After two delay periods ($t = 12$ year) this is 14%. This corresponds with the theoretical values for exponential decay. The smaller the time step chosen in the model, the more accurate the figure of the outflow value "natural degradation rate" will be.

Since the half time is 4.2 years (see (11.6)), the outflow value of $t = 8$ [year] is approximately half of the value of $t = 4$ [year] and the outflow value of $t = 12$ [year] is approximately half of the value of $t = 8$ [year].

Depending on the type of variable for the inflow rate, the dynamic behavior for the stock material in transit and the outflow rate will differ for a system with first-order material delay. We demonstrate this by adding two fictive theoretical new types of behavior over time for the parameter OIL SPILL. A sinusoidal input is chosen for OIL SPILL. We introduce the shadow variable *<Time>* in Vensim® (see Fig. 11.15) to model the variable OIL SPILL = 5e+07*SIN((2 * 3.14159/100) *Time). Next, we opt for a step change increase of 50,000,000 at time $t = 0$. In this case, the constant "OIL SPILL" is modeled in Vensim® as OIL SPILL = STEP (5e+07, 0) [Gallon/year].

Figure 11.16 shows the difference in behavior of the inflow rate variables for an inflow rate of a pulse with height 5e+07 at $t = 0$ year and a duration of one time step

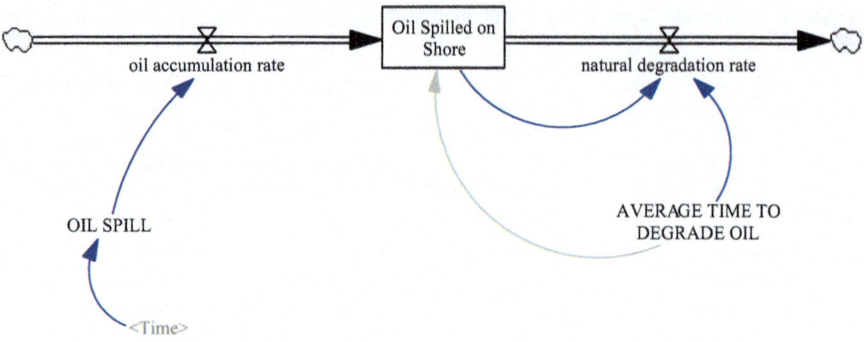

Fig. 11.15 Identical to Fig. 11.7 but shadow variable *<Time>* has been added

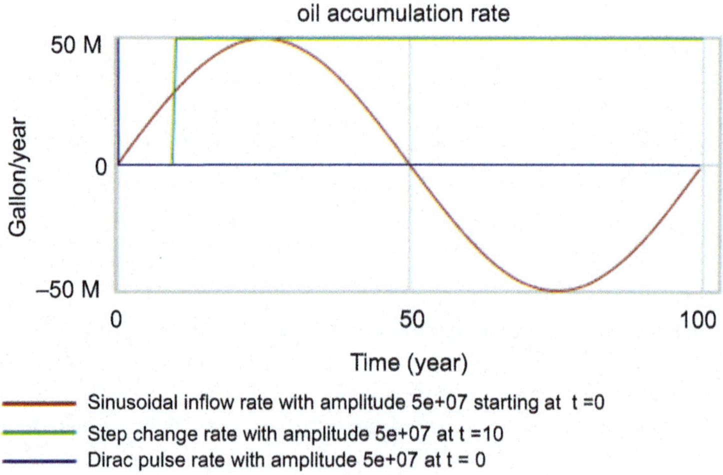

Fig. 11.16 Different types of inflow rates for the Oil Spill example

(in this case 1/16 year), a step change increase with height 5e+07 at $t = 10$ year, and a sinusoidal input rate with an amplitude of 5e+07 Gallon starting at $t = 0$ year.

The state of the stock "Oil Spilled on Shore," which is the Material in Transit, is depicted in Fig. 11.17 for the different input variables. The fictive case of the sinusoidal input cannot be realized physically. However, the system's response to different types of inflow rates is instructive to understand the dynamic behavior of the first-order material delay system. The step increase input will lead to an increase in the stock "Oil Spilled on Shore" in the first years. After a transition period, a steady state behavior of the stock "Oil Spilled on Shore" will become apparent. The AVERAGE TIME TO DEGRADE OIL delays the state change in "Oil Spilled on Shore" (Fig. 11.17) and the outflow rate (Fig. 11.18).

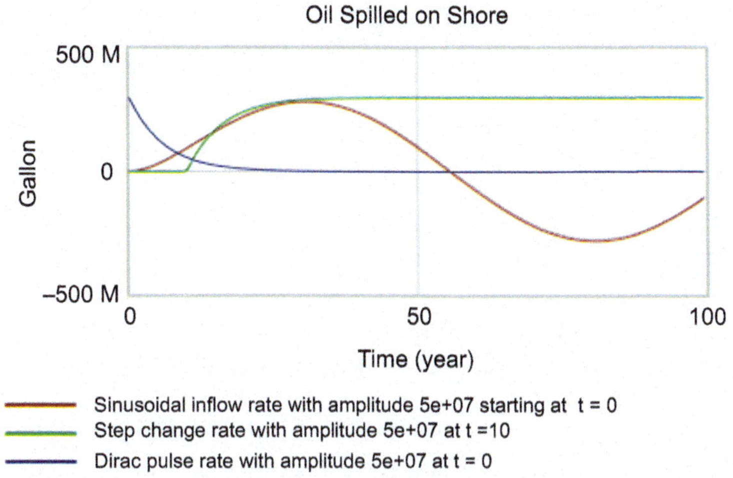

Fig. 11.17 Different types of stock behavior related to the different types of inflow behavior for the Oil Spill example

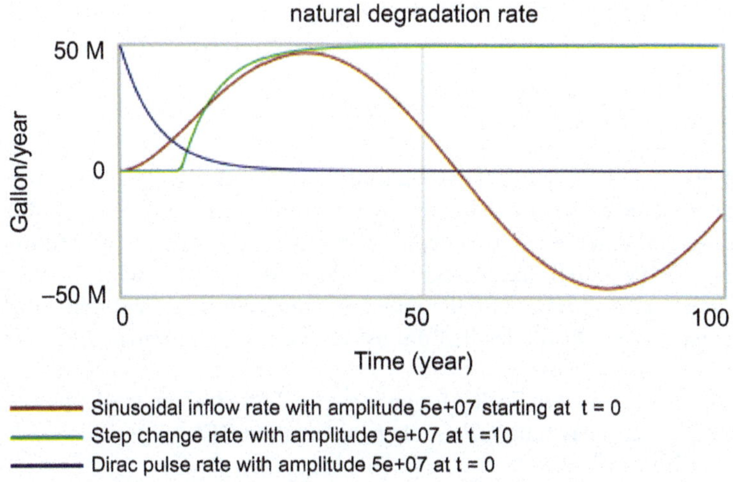

Fig. 11.18 Different types of outflow rates for the Oil Spill example

4 N-th Order Material Delay

An n-th or higher-order material delay repeats the first-order material delay n times. The outcome of each stage in a first-order material delay is used as an input for the next first-order material delay, as depicted in Fig. 11.19 for a third-order material delay ($n = 3$).

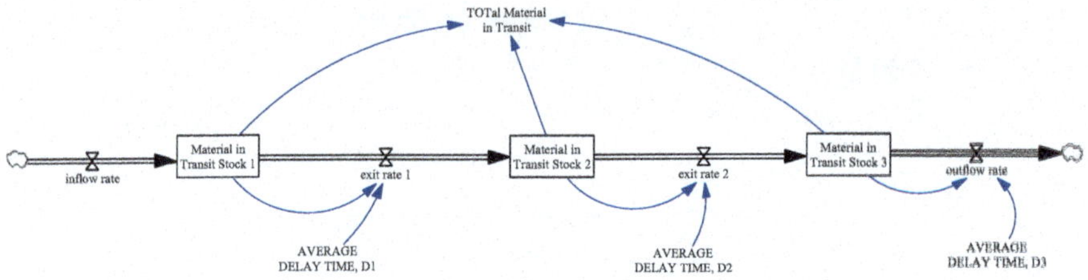

Fig. 11.19 Stock and flow diagram of a third-order material delay

Note that in this example of a third-order material delay, the system contains three stocks. These stocks are linked with flows and average delay times as follows:

$$\text{exit rate 1} = \text{Material in Transit Stock 1/AVERAGE DELAY TIME } D1$$

$$\text{exit rate 2} = \text{Material in Transit Stock 2/AVERAGE DELAY TIME } D2$$

$$\text{Outflow Rate} = \text{Material in Transit Stock 3/AVERAGE DELAY TIME } D3$$

The sum of the Material in Transit for the entire system is captured in the auxiliary variable TOTal Material in Transit

$$\text{TOTal Material in Transit} = \sum_{i=1}^{n} \text{Material in Transit Stock } i \tag{11.7}$$

The response of a first, second and third-order material delay system, with an average time delay of 10 units for each outflow rate on a Dirac pulse inflow depositing 100 units at $t = 0$ is depicted in Fig. 11.20. The outflow from the first stock is an exponential decay. The initial slope for the outflow of the second and third stock equals zero because the initial input to the second and third stock of the delay chain is zero. For a detailed discussion we refer to Sterman (2000) and MIT (1999).

If a third-order material delay for an input "*in*" for all three stocks in transit uses the same average delay time "*dtime*" then the function DELAY3 (*in_* , *_dtime_*) can be used in Vensim®.

5 Information Delays

In addition to physical material delays, there are also delays in our perception, mindset, and attitude. These delays are called *information delays*. They are typical delays that exist in information feedback channels. Similar to material delays, information delays are characterized by the order of the delay and by the average length of the delay. The main difference with material delays is that information

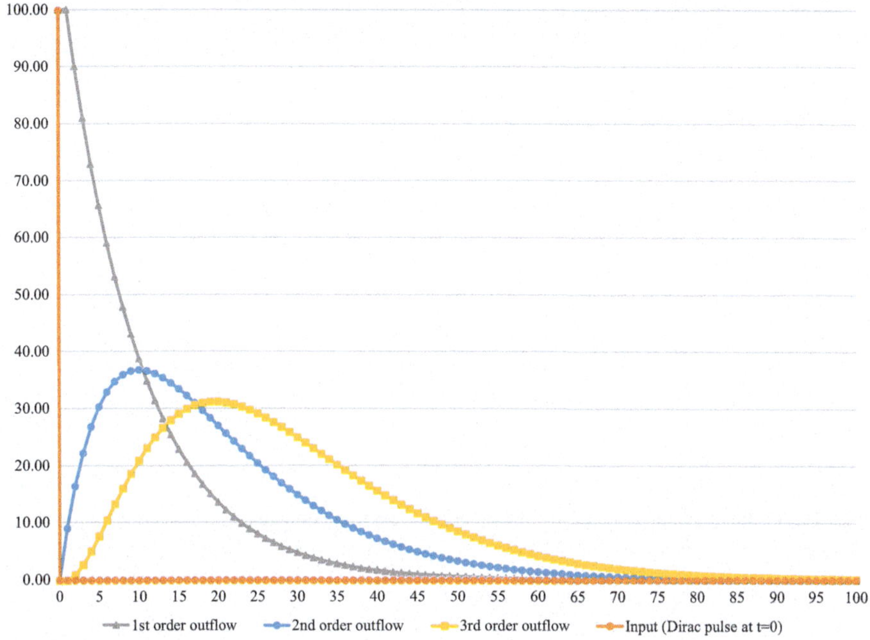

Fig. 11.20 Dirac pulse inflow rate outflow response for a first, second and third-order system

delays have no physical inflow to the stock of material in transit. A mindset or the state of mind often changes slowly. Consider a situation where you had to change your attitude. It probably took weeks or even months before the change was effectively in place, and the change was probably gradually.

We first consider the simplest information delay, which is also known as *exponential smoothing* or *adaptive expectations* (Sterman 2000). The typical feedback structure of this information delay is depicted in Fig. 11.21.

In adaptive expectations, the change in the state of mindset or adaptive expectation is called the *Perceived Value* $\widehat{O}$; it is represented as a stock. The change in the perceived value is proportional to the gap between the current value of the input O and the perceived value $\widehat{O}$ divided by the average adjustment time AT:

$$\text{change in perceived value} = \frac{\left(O - \widehat{O}\right)}{AT} \tag{11.8}$$

The change in perceived value is a flow that represents the step-by-step change in mindset over time driven by the gap, called the "ADAptation gap", between the current belief, called the "CURrent value O", and the actual value of the new state of belief represented by the stock "Perceived Value $\widehat{O}$".

The output of this system is similar to that of a first-order linear negative feedback system. Therefore, it demonstrates goal-seeking behavior, as depicted in Fig. 11.22.

The change in perceived value is proportional to (11.8), as depicted in Fig. 11.23.

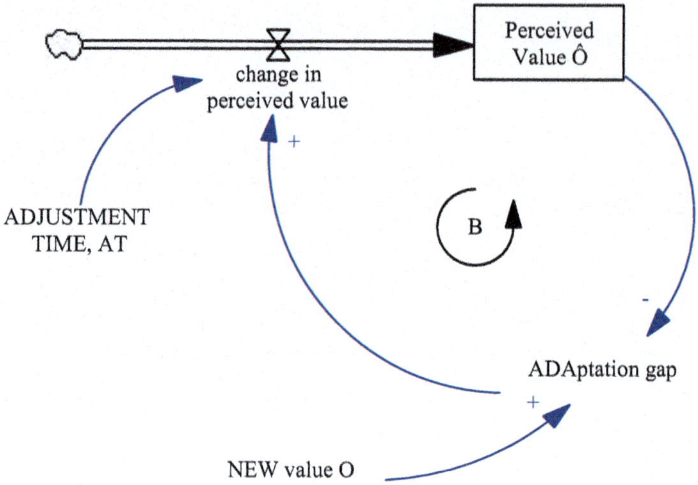

Fig. 11.21 Stock and flow diagram of exponential smoothing or adaptive expectations information delay

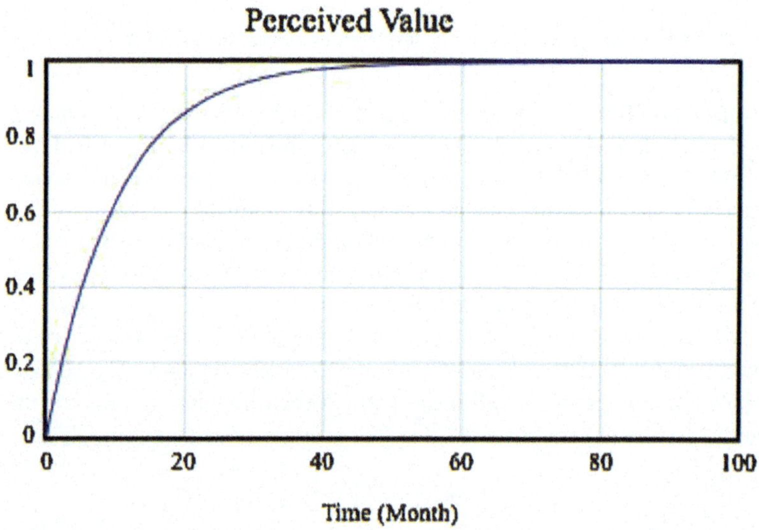

Fig. 11.22 Response of the Perceived Value $\widehat{O}$ to a step change in the input NEW value at $t = 0$ from 0 to 1

We will illustrate first-order information delay with an example in the next section.

Vensim® provides the functions DELAY INFORMATION, SMOOTH3, and SMOOTH3I to support the modeling of information delays.

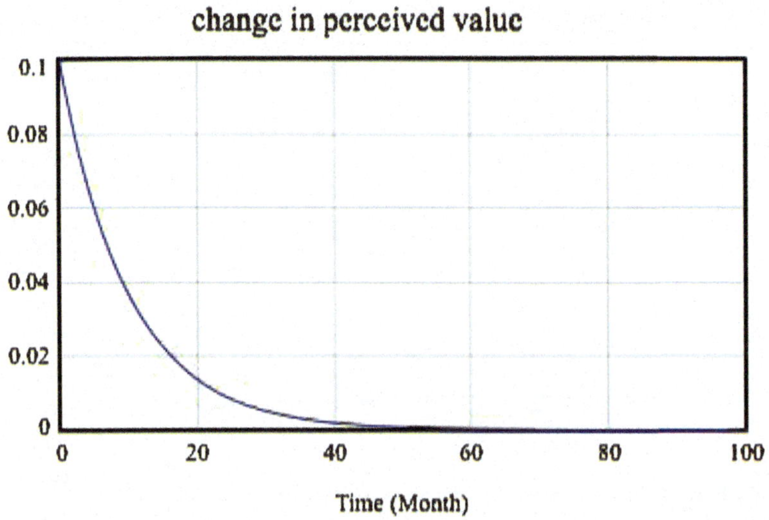

Fig. 11.23 Change in perceived value as a response to a step change in the input NEW value at $t = 0$ from 0 to 1

6 Dynamics of Oscillating Behavior

Systems may show oscillating behavior if significant time delays are present in the negative feedback loop of the system AND if decision makers fail to account for these delays (Sterman 2000). As discussed in Chap. 9, delays in the negative feedback loop may occur in several places in the negative or balancing feedback loop.

An illustration of oscillating behavior is what sometimes happens when we take a hot shower. If the water temperature does not increase as fast as we expect, we increase the temperature setpoint in some way. This can cause the expected temperature to be reached faster. Consequently, the actual shower temperature will exceed the desired temperature after some time. When this happens, we tend to decrease the setpoint. These adjustments continue for a while until the desired temperature is reached.

A stock and flow diagram of the abovementioned system of controlling the temperature in the shower is depicted in Fig. 11.24. The model consists of two state variables: (i) "Shower Temperature" represents the temperature of the water in the shower controlled by a thermometer, and (ii) a mental state variable called "Desired Temperature". For this example, we assume that the perception time delay is smaller than the physical adjustment time required to control the hot water in the shower. As a consequence, the "DESired temperature gap" = "SETPOINT TEMPERATURE" – "Desired Temperature" will differ from the physical "TEMperature gap" = "SETPOINT TEMPERATURE" – "Shower Temperature". This leads to a "PERception gap" = "Desired Temperature" – "Shower Temperature".

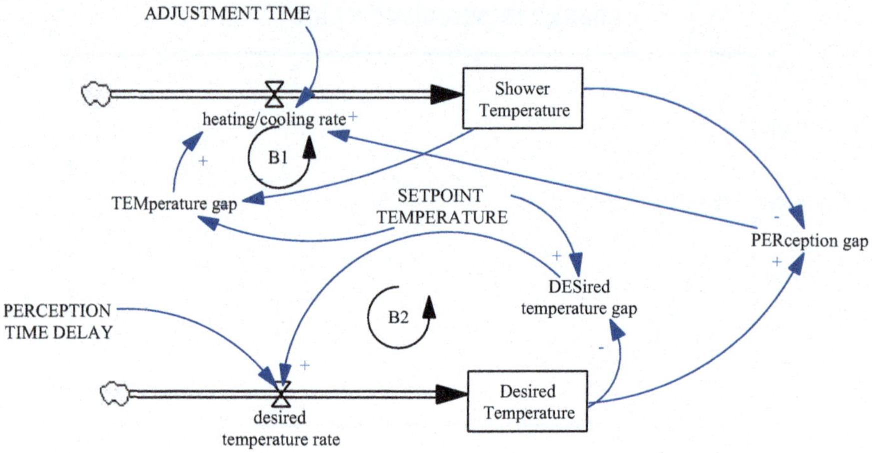

Fig. 11.24 Stock and flow diagram of the Shower Temperature control case

To show how oscillating behavior may occur, we use this perception gap as a multiplier for the adjustment of the shower's thermostat by multiplying this factor with the inflow of the shower's thermostat called the "heating/cooling rate" = ("TEMperature Gap"/"ADJUSTMENT TIME")* "PERception gap". As a result, the setpoint of the shower will be unstable and will oscillate. Figure 11.24 shows the stock and flow diagram of the shower temperature control system.

The Vensim® program code for this temperature-controlling example with oscillating behavior looks like this:

(01) ADJUSTMENT TIME = 10
Units: Second

(02) Desired Temperature = INTEG (desired temperature rate, 20)
Units: °C

(03) DESired temperature gap = SETPOINT TEMPERATURE-Desired Temperature
Units: °C

(04) desired temperature rate = DESired temperature gap/PERCEPTION TIME DELAY
Units: °C/Second

(05) FINAL TIME = 100
Units: Second
The final time for the simulation.

(06) "heating/cooling rate" = (TEMperature gap/ADJUSTMENT TIME)*PERception gap
Units: °C/Second

(07) INITIAL TIME = 0
Units: Second
The initial time for the simulation.

(08) PERception gap = Desired Temperature-Shower Temperature
Units: °C

(09) PERCEPTION TIME DELAY = 4
Units: Second

(10) SAVEPER = TIME STEP
Units: Second [0,?]
The frequency with which output is stored.

(11) SETPOINT TEMPERATURE = 45
Units: °C

(12) Shower Temperature = INTEG ("heating/cooling rate", 20)
Units: °C

(13) TEMperature gap = SETPOINT TEMPERATURE-Shower Temperature
Units: °C

(14) TIME STEP = 1
Units: Second [0,?]
The time step for the simulation.

The desired temperature response that forms the mental state of the expected water temperature in the shower over time is depicted in Fig. 11.25.

The physical increase in water temperature lags in the mindset of the person taking a shower because the PERCEPTION TIME DELAY is 4 s (i.e., the shower temperature increase from 20 °C to 45 °C is expected to be realized for 99% in 4*5 = 20 s for a system showing goal-seeking behavior based on Table 10.1), and the ADJUSTMENT TIME of the thermostat is 10 s (i.e., it takes about 10*5 = 50 s to realize a shower temperature of 45 °C for 99%). This mentally perceived delay in increasing the water temperature is expressed by the perception gap that is defined as the difference between the "Shower Temperature" and the "Desired Temperature". This perception gap is depicted in Fig. 11.26. If the state of the "Shower Temperature" would evolve more or less equally to the state of the "Desired Temperature", then this difference would tend towards zero and would not influence the heating/cooling behavior any longer. This finally takes place once the delays are no longer dominant.

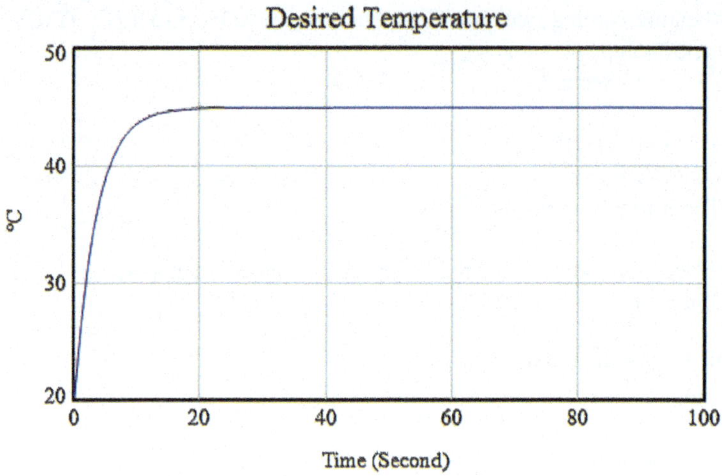

Fig. 11.25 Desired temperature response

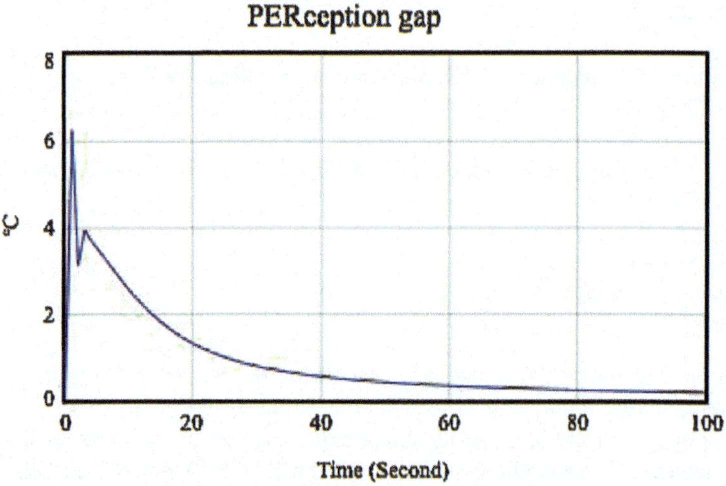

Fig. 11.26 Perception gap in heating up of the shower water temperature

The perception gap influences the heating of the water temperature. Figure 11.26 shows how the setpoint is adjusted a few times (first it is perceived as too cold, then too hot), resulting in an oscillating heating/cooling rate (Fig. 11.27).

The causal strip diagram in Fig. 11.28 summarizes the root cause of the oscillating behavior of the heating/cooling rate.

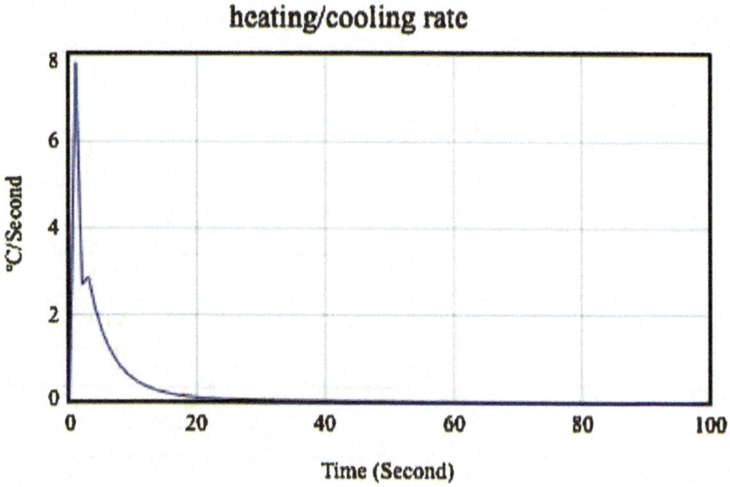

Fig. 11.27 Oscillating heating/cooling rate because of the temperature perception gap

7 Dynamics of Limits to Growth

The Limits to Growth archetypical behavior shows that reinforcing feedback loops push a system beyond its sustainable limits. Once this happens, balancing feedback loops will dominate the system's behavior in order to prevent its collapse. We demonstrate this behavior using a case related to a case described by Mirchi et al. (2012) that describes the dynamic behavior of an agricultural system using groundwater as a source of irrigation water. Agricultural growth in this system will increase the demand for groundwater. Agricultural growth is profitable at first, and farmers keep pumping up more water to irrigate the land. At this moment, the reinforcing loop, *R*, in Fig. 11.29 is dominant. However, pumping up water will increase the price of water because the groundwater level will drop, driven by continuous agricultural growth. At a certain moment, water pumping costs will rise dramatically, and farmers will no longer make more profit by extending their business. That is when the balancing control loop, *B*, in Fig. 11.29 becomes dominant. Note that in a very extreme case, the non-renewable groundwater resource may be completely depleted if farmers do not stop expanding.

Figure 11.30 depicts a causal loop diagram for the case of two farmers A and B pumping up the same amount of water. A reinforcing loop, respectively *R1* and *R2*, drives the additional groundwater usage of farmers A and B to increase the net profit of each farmer. This depletes the commonly used groundwater and increases the pumping costs for farmers A and B. As a result, there is a limit to growing the net profit for farmers A and B that is captured in the balancing feedback loops *B1* and *B2*.

The corresponding stock and flow chart of this system is depicted in Fig. 11.31.

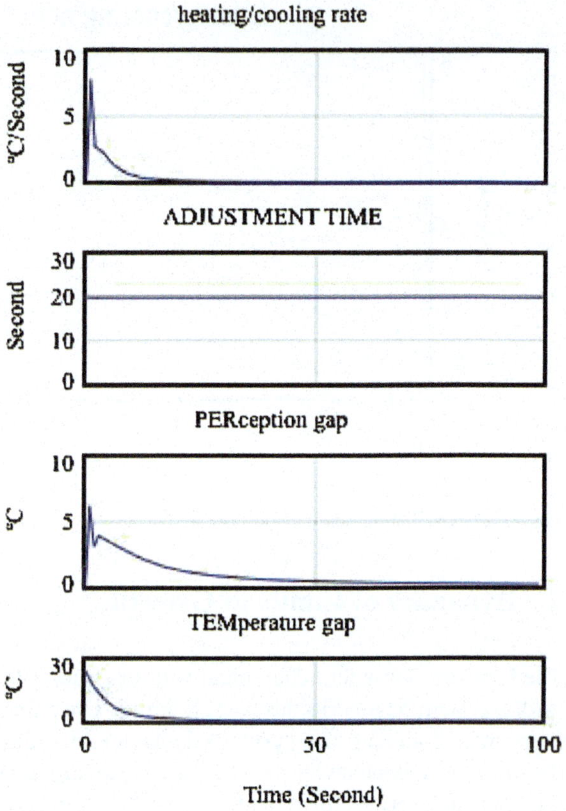

Fig. 11.28 Causal strip diagram for the oscillating behavior of the heating/cooling rate

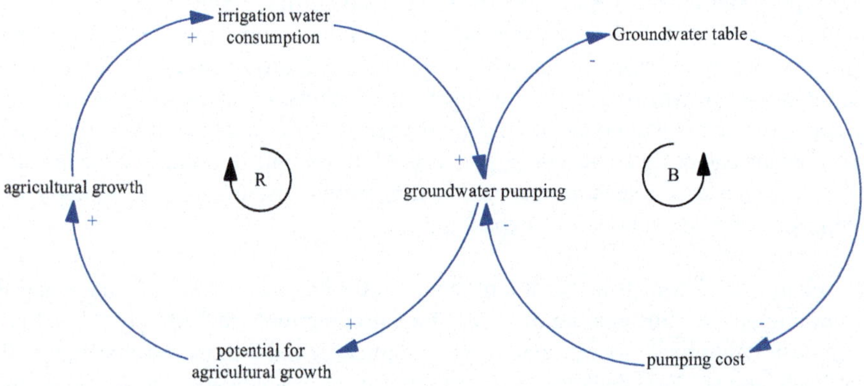

Fig. 11.29 Causal loop diagram of pumping up water by farmers adapted from Mirchi et al. (2012)

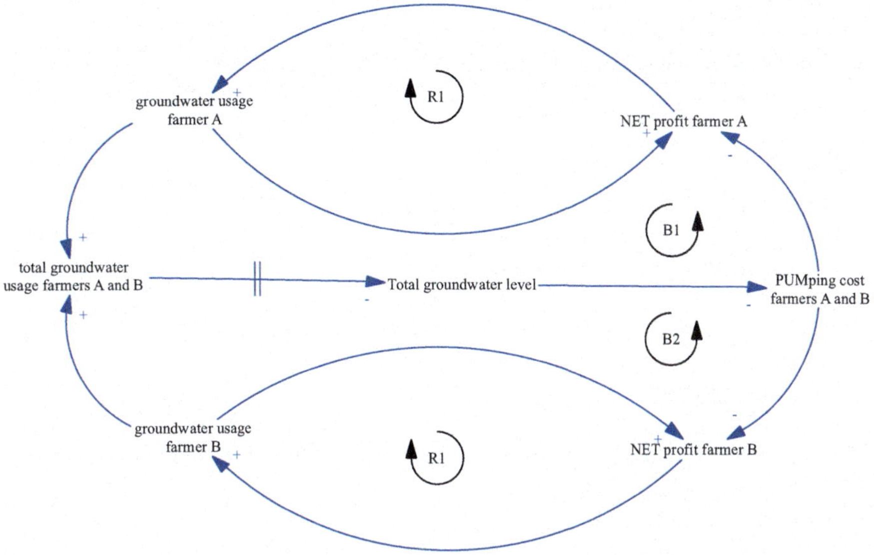

Fig. 11.30 Causal loop diagram depicting farmers A and B pumping up the same amount of water. (Adapted from Mirchi et al. 2012)

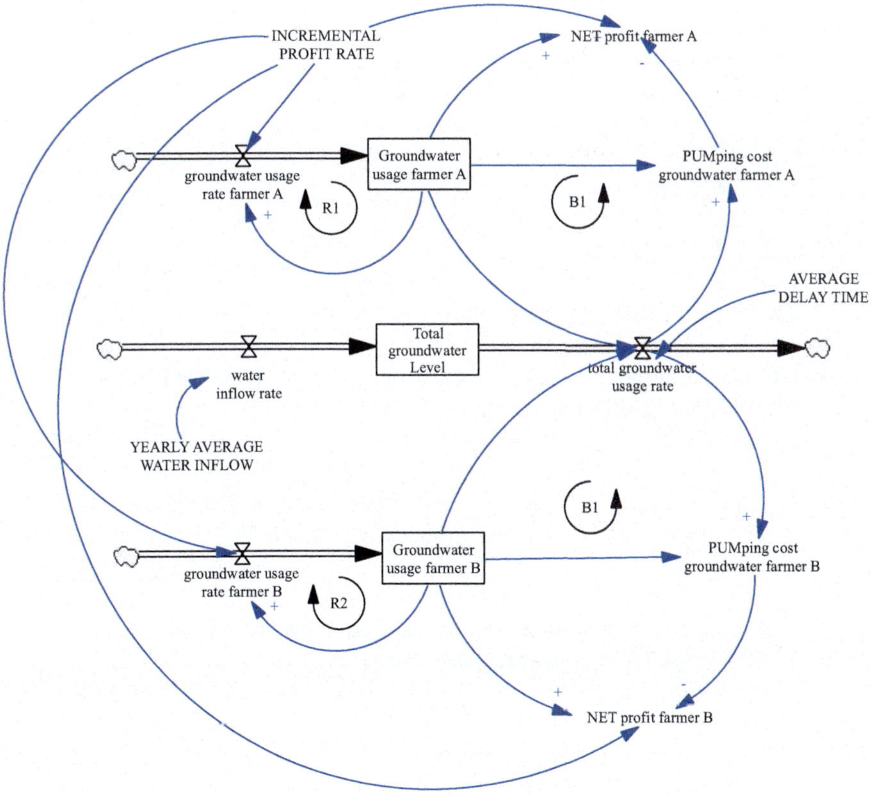

Fig. 11.31 Stock and flow diagram of the water pumping example

The Vensim® program code for this problem looks like this:

(01) AVERAGE DELAY TIME = 1
Units: Year

(02) FINAL TIME = 100
Units: Year
The final time for the simulation.

(03) Groundwater usage farmer A = INTEG (groundwater usage rate farmer A, 6000)
Units: m^3/Year

(04) Groundwater usage farmer B = INTEG (groundwater usage rate farmer B, 6000)
Units: m^3/Year

(05) groundwater usage rate farmer A = Groundwater usage farmer A*INCRE MENTAL PROFIT RATE
Units: m^3/Year

(06) groundwater usage rate farmer B = Groundwater usage farmer B*INCRE MENTAL PROFIT RATE
Units: m^3/Year

(07) INCREMENTAL PROFIT RATE = 0.02
Units: KEuro/m^3

(08) INITIAL TIME = 0
Units: Year
The initial time for the simulation.

(09) NET profit farmer A = Groundwater usage farmer A*INCREMENTAL PROFIT RATE-PUMping cost groundwater farmer A
Units: KEuro

(10) NET profit farmer B = Groundwater usage farmer B*INCREMENTAL PROFIT RATE-PUMping cost groundwater farmer B
Units: KEuro

(11) PUMping cost groundwater farmer A = (total groundwater usage rate/ 16,000)*SQRT(Groundwater usage farmer A)
Units: KEuro

(12) PUMping cost groundwater farmer B = (total groundwater usage rate/ 16,000)*SQRT(Groundwater usage farmer B)
Units: KEuro

(13) SAVEPER = TIME STEP
Units: Year [0,?]
The frequency with which output is stored.

(14) TIME STEP = 0.03125
Units: Year [0,?]
The time step for the simulation.

(15) Total groundwater Level = INTEG (water inflow rate-total groundwater usage rate, 100,000)
Units: m^3

(16) total groundwater usage rate = (Groundwater usage farmer A + Groundwater usage farmer B)/AVERAGE DELAY TIME
Units: m^3/Year

(17) water inflow rate = YEARLY AVERAGE WATER INFLOW
Units: m^3/Year

(18) YEARLY AVERAGE WATER INFLOW = 8000
Units: m^3/Year

We suppose that rainfall provides a constant yearly inflow rate of 8000 m^3 of water per year (i.e., 8,000,000 liters of water per year) on the land used by farmers A and B. This water inflow rate is the part of the rainfall that infiltrates the soil and replenishes the groundwater level (Fig. 11.32).

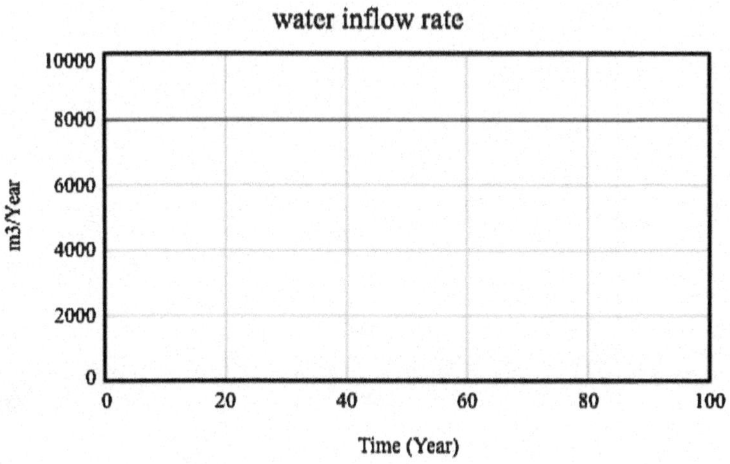

Fig. 11.32 Yearly water inflow rate on the land of farmers A and B

Since the rainfall is not sufficient, farmers A and B have to irrigate the land and pump up groundwater. The more land they use for agriculture, the more groundwater they will need (Fig. 11.33).

This increased use of groundwater gradually depletes the stock of groundwater. A negative stock in Fig. 11.34 denotes more consumption of groundwater than the yearly water inflow can compensate.

The extra water groundwater usage generates a benefit for farmers A and B and an additional pumping cost. We assume that at time $t = 0$, each farmer needs 6000 m^3 of

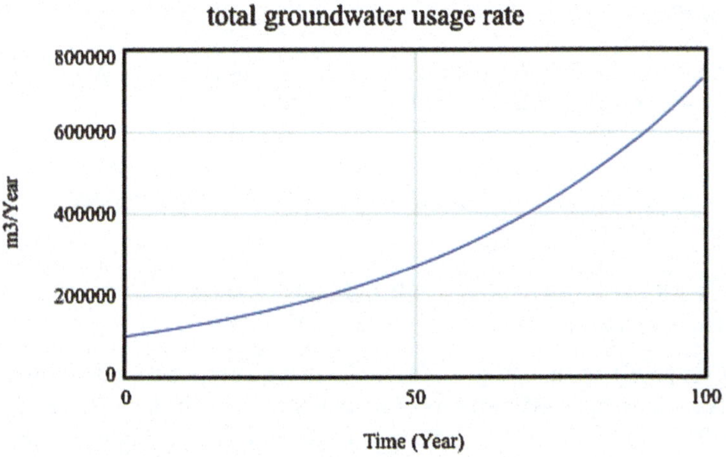

Fig. 11.33 Total groundwater usage by farmers A and B over time

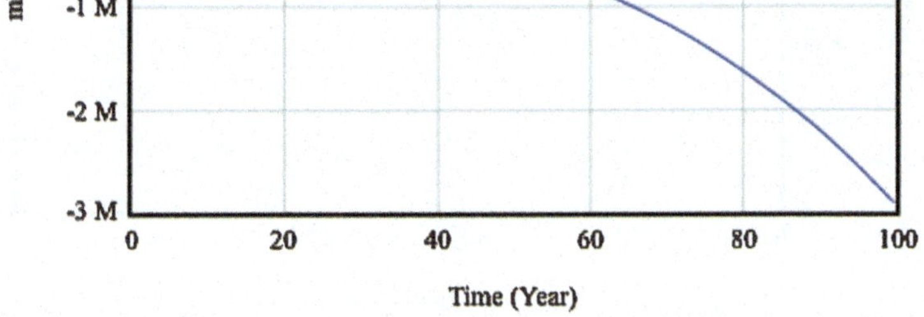

Fig. 11.34 Total groundwater usage rate

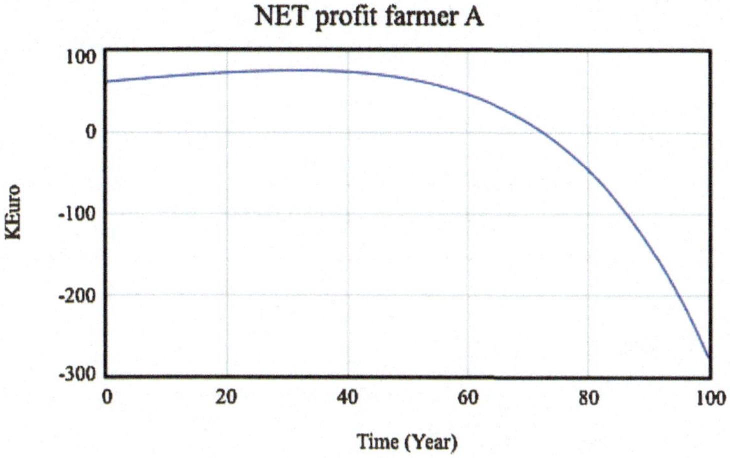

Fig. 11.35 Net profit evolution for farmer A pumping up more groundwater to expand his business

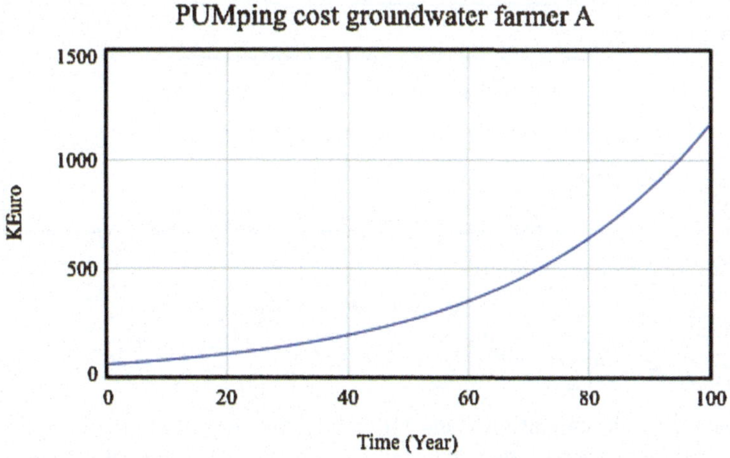

Fig. 11.36 Increased pumping cost for farmer A due to the expansion of his business

water per year. The additional net profit of expanding the agricultural activities for each farmer is shown in Fig. 11.35 for farmer A. During the first years, the net profit grows each year before it gradually drops and then becomes negative and therefore will be no longer profitable.

The root cause of the ultimately decreasing net profit is the increased groundwater pumping cost. The lower the groundwater level, the more energy is required to pump up the groundwater. Over time, the groundwater will be situated at ever lower levels. This results in an increased pumping cost that cannot be compensated with an increased incremental profit rate of selling more crops. We assume that the incremental profit rate stays steady over the entire examined period of 100 years (Fig. 11.36).

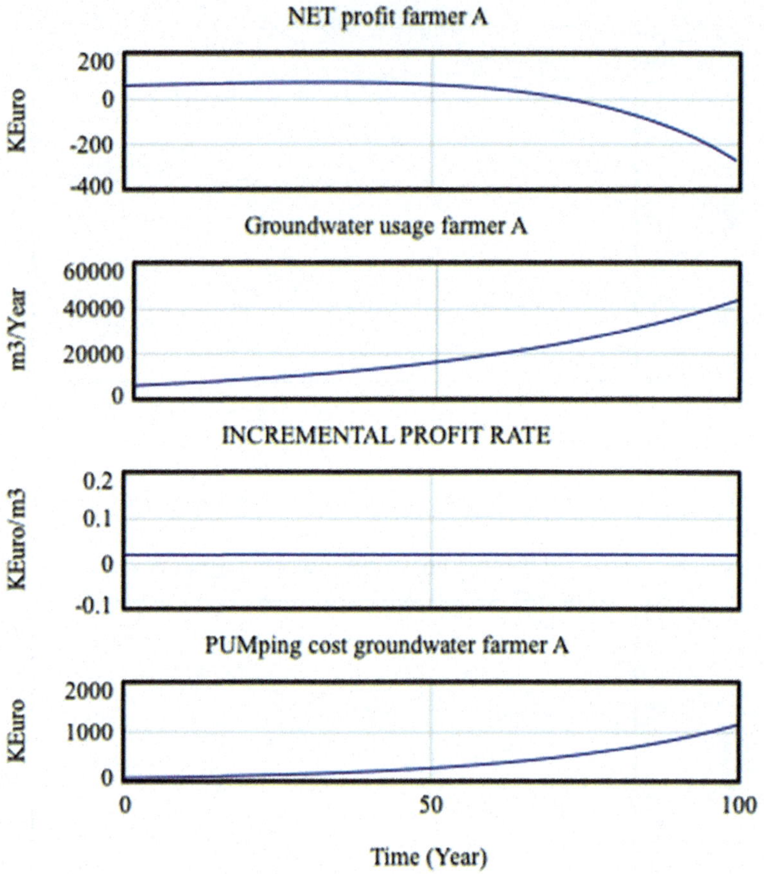

Fig. 11.37 Causal strip net profit of farmer A

In summary, the expansion of the agricultural business of farmers A and B will be limited due to pumping costs that will exceed the net profit associated with the growth of their agricultural business. This is shown in the causal strip in Fig. 11.37 for farmer A. The same rationale is applicable for farmer B.

Chapter 12
Nonlinear Behavior

The behavior of many real-life systems can be described using nonlinear functions. Therefore, nonlinear behavior and its modeling are of major importance in system dynamics. In this chapter, we discuss four important functions that are commonly used to model nonlinear behavior: the table function, MIN and MAX functions, and the IF...THEN, ELSE function. Many more nonlinear functions are available in Vensim® and are waiting to be explored.

A system is said to behave nonlinearly if the change of the output is not proportional to the change of the input. By contrast, a system is called linear if it can be described by a function to which the following is applicable: $f(\alpha \cdot x + \beta \cdot y) = \alpha \cdot f(x) + \beta \cdot f(y)$. Where α and β represent real values. In all other cases, the functions are called nonlinear. Take, for example, the function $f(x) = x^2$. This parabolic function is nonlinear because $(2 \cdot x)^2 \neq 2 \cdot x^2$.

1 Table Function

A table function allows us to represent any relationship defined by known pairs of points (x_i, y_i) of this function. It allows to model linear and nonlinear functions. The table effect of variable X on variable Y is generally represented by the points (x_1, y_1), (x_2, y_2), ..., (x_n, y_n). It can be a representation of the monitoring of values in a process without knowing how this process should be described mathematically.

In Vensim®, the table function is represented by the function LOOKUP. An example in Sect. 4 explains how this function works.

D. Inghels, *Introduction to Modeling Sustainable Development in Business Processes*, https://doi.org/10.1007/978-3-030-58422-1_12

2 MIN and MAX

The MIN and MAX functions return, respectively, the smaller and larger value of the values that are listed between the brackets following this function.

Take $C = \text{MIN}\ (5, 8)$ as an example. The outcome of this function will be $C = 5$. The figures between the brackets can be replaced with variables such as A and B. $C = \text{MIN}\ (A, B)$ returns the smaller values of the current state of the variables A and B. An example in Sect. 4 will demonstrate how this function is used.

3 IF...Then, Else

The IF...Then, Else function $D = \text{IF}\ A$ Then B ELSE C should be noted in Vensim® syntax as D = IF THEN ELSE (A, B, C). Logical statements such as $A = F > E$ are often inserted. If the logical result for A is true, D gets the value B assigned. If not, D gets the value C assigned. A, B and C can be constants or variables. This function is also demonstrated in the example in the next section.

4 Caste Study: My Domestic Power Consumption

We demonstrate the above-mentioned nonlinear functions using the case study of my domestic power consumption in 2017. My domestic electric power is produced using solar panels. If they do not generate enough electric power to be self-sufficient, power is consumed from the grid. My home is heated using natural gas from the grid.

The gas and electric power consumption are recorded at the end of each month. The 2017 records are depicted in Table 12.1. If the solar panels produced more electric power than we consumed in a given month, the resulting monthly consumption of electric power is denoted with a negative number. The excessive electric power was fed back into the grid.

Solar panels should predominantly cover the electric power that my family needs. Because we invest in low-energy consuming devices every year, we have noticed a negative year-over-year electric power consumption at the end of each year. In other words, we produce more electric power than we consume over an entire year. The amount of excessive power produced by our solar panels in 2017 was 201 kWh. Excessive power is not reimbursed by the grid operator. The electric power that my solar panels produce but that we do not consume by the end of the year is, financially speaking, lost. Therefore, it is worthwhile examining whether we could use this electric power as an energy source to heat the bathroom in the winter using a new electric-powered radiator. To be on the safe side, we will only start using electric power for heating once we have collected excessive power during the running year. We take an average electric power consumption of 170 kWh/month into account.

Table 12.1 My monthly domestic gas and electric power consumption in 2017

Year	Month	Monthly gas consumption [m^3]	Monthly electric power consumption [kWh]	Production solar energy [kWh]
2017	Jan	499	99	83
	Feb	331	92	76
	Mar	254	−39	210
	Apr	209	−65	219
	May	150	−212	336
	Jun	19	−149	295
	Jul	20	−129	289
	Aug	20	−63	236
	Sep	45	−88	197
	Oct	146	50	129
	Nov	362	130	69
	Dec	414	173	22
Total			−201	

Moreover, electric power will only be consumed in the winter months when the monthly gas consumption is above 100 m^3/month.

This problem is modeled in the stock and flow diagram of Fig. 12.1.

The inputs for this model are the monthly gas consumption and solar power production of 2017 (see Table 12.1). The monthly consumption rates listed in Table 12.1 are inserted in the Vensim® model using the LOOKUP function. Take the monthly solar power production as an example. The monthly amount of solar power produced is recorded in the last column of Table 12.1. These values have to be added in a lookup table per pair (x_i, y_i) of which x_i represents the months of year 2017 (0 = January 1, 1 = February 1, etc.) and y_i represents the solar power produced in the previous month and recorded on the first day of the month (e.g., Recording on February 1: $x_i = 1$ and $y_i = 83$ kWh). These numbers are transferred into the model in the variable "SOLAR POWER PRODUCTION LOOKUP". The numbers can be filled in by selecting the type of variable 'Lookup' and then clicking the button "As Graph" (Fig. 12.2). The graphical representation of the lookup table is shown in Fig. 12.3.

The monthly gas consumption records are also inserted in the model. The monthly amount of gas consumption is recorded in the third column of Table 12.1. These numbers are input into the model in the variable "GAS CONSUMPTION LOOKUP" in the same way (Fig. 12.4).

The records of the total solar production and gas consumption over all the months of the year are included in the SD model using the respective stock variables "Total solar power production" and "Total gas consumption".

The constant "MONTHLY AVERAGE ELECTRIC POWER NEED" represents the monthly projected average power need of 170 kWh for our household. The calculated projected accumulative average monthly power need is assigned to the auxiliary variable "CUMulative monthly estimated electric power consumption". To

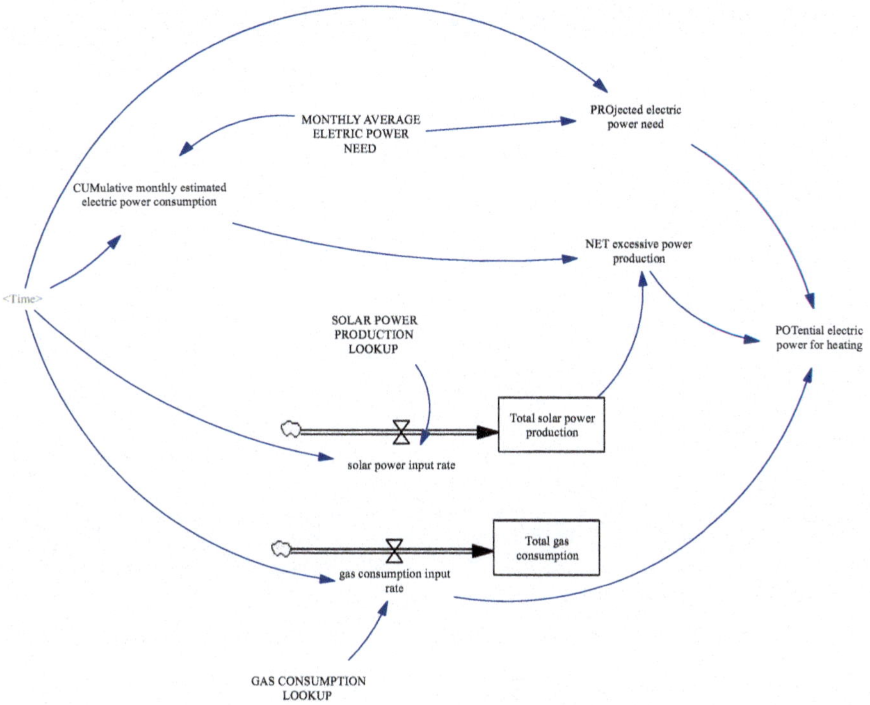

Fig. 12.1 Stock and flow diagram of my domestic annual energy consumption in 2017

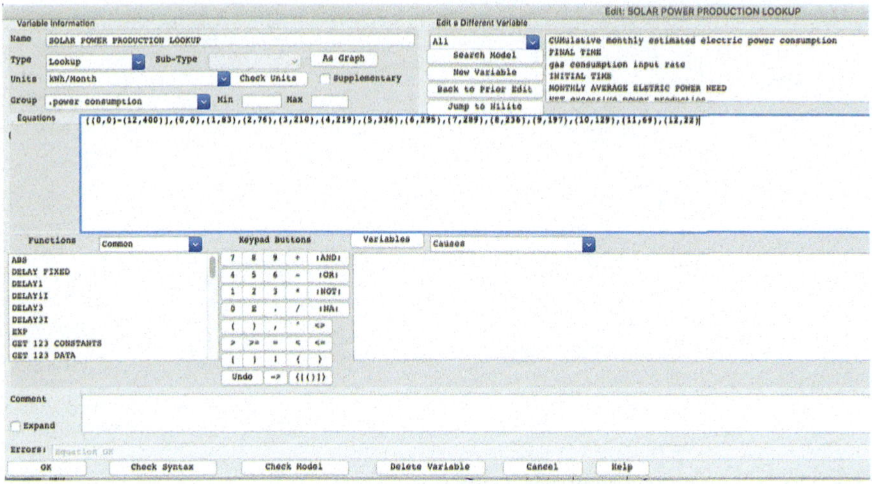

Fig. 12.2 The SOLAR POWER PRODUCTION LOOKUP variable

Fig. 12.3 The graphical representation of the SOLAR POWER PRODUCTION LOOKUP variable

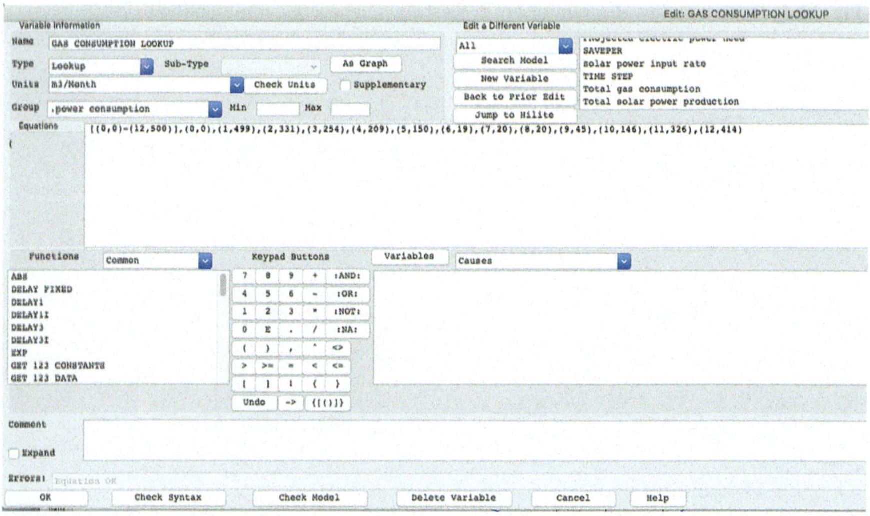

Fig. 12.4 The GAS CONSUMPTION LOOKUP variable

add up the projected need for each month, we make use of the shadow variable <Time>. The selected Time Step is 1 month.

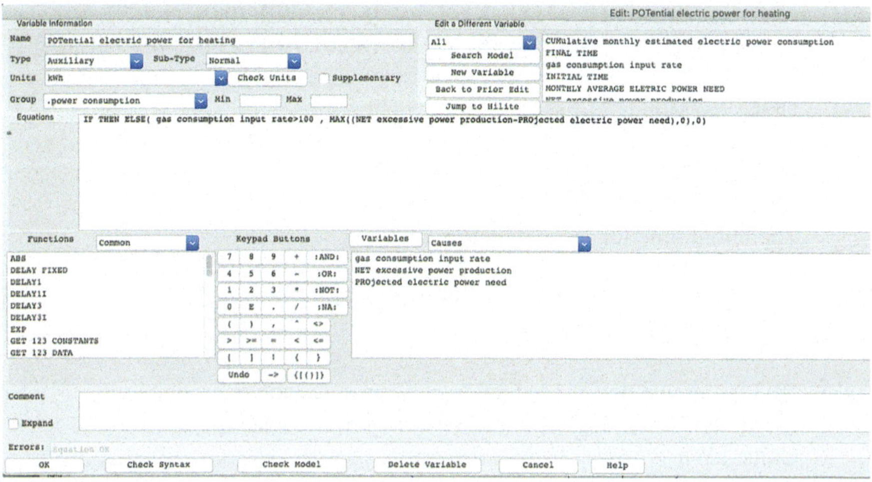

Fig. 12.5 IF…THEN, ELSE expression for the potential electric power production used for electrical heating

In order to assess whether the excess of solar power can be used for electric heating, we first calculate the remaining electric power needed for the rest of the year. To this end, the variable "PROjected electric power need" is used.

This variable is used as an input to compute the "POTential electric power for heating". The decision rules that were discussed earlier apply: the excessive solar power production that can be used for electric heating is the difference between the "Net excessive power production" and the "PROjected electric power need" if the "gas consumption input rate" exceeds 100 m^3/month and the aforementioned difference is positive. This is modeled using the IF…THEN ELSE function for the auxiliary variable "POTential electric power for heating" as depicted in Fig. 12.5.

Taking all the previous information into account, the Vensim® code for this model looks like this:

(01) CUMulative monthly estimated electric power consumption = MONTHLY AVERAGE ELECTRIC POWER NEED*Time
Units: kWh

(02) FINAL TIME = 12
Units: Month
The final time for the simulation.

(03) gas consumption input rate = GAS CONSUMPTION LOOKUP (Time)
Units: m^3/Month

(04) GAS CONSUMPTION LOOKUP ([(0,0)-(12,500)], (0,0), (1,499), (2,331), (3,254), (4,209), (5,150), (6,19), (7,20), (8,20), (9,45), (10,146), (11,326), (12,414))
Units: m^3/Month

(05) INITIAL TIME = 0
Units: Month
The initial time for the simulation.

(06) MONTHLY AVERAGE ELECTRIC POWER NEED = 170
Units: kWh/Month

(07) NET excessive power production = Total solar power production-CUMulative monthly estimated electric power consumption
Units: kWh

(08) POTential electric power for heating = IF THEN ELSE (gas consumption input rate > 100, MAX ((NET excessive power production – PROjected electric power need),0),0)
Units: kWh

(09) PROjected electric power need = (12-Time)*MONTHLY AVERAGE ELECTRIC POWER NEED
Units: kWh

(10) SAVEPER = TIME STEP
Units: Month [0,?]
The frequency with which output is stored.

(11) solar power input rate = SOLAR POWER PRODUCTION LOOKUP(Time)
Units: kWh/Month

(12) SOLAR POWER PRODUCTION LOOKUP([(0,0)- (12,400)], (0,0), (1,83), (2,76), (3,210), (4,219), (5,336), (6,295), (7,289), (8,236), (9,197), (10,129), (11,69), (12,22))
Units: kWh/Month

(13) TIME STEP = 0.03125
Units: Month [0,?]
The time step for the simulation.

(14) Total gas consumption = INTEG (gas consumption input rate, 0)
Units: m^3

(15) Total solar power production = INTEG (solar power input rate, 0)
Units: kWh

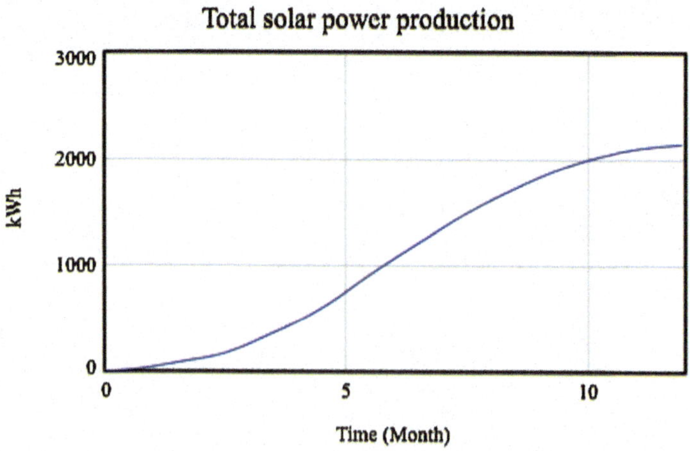

Fig. 12.6 The accumulated monthly solar power production in 2017

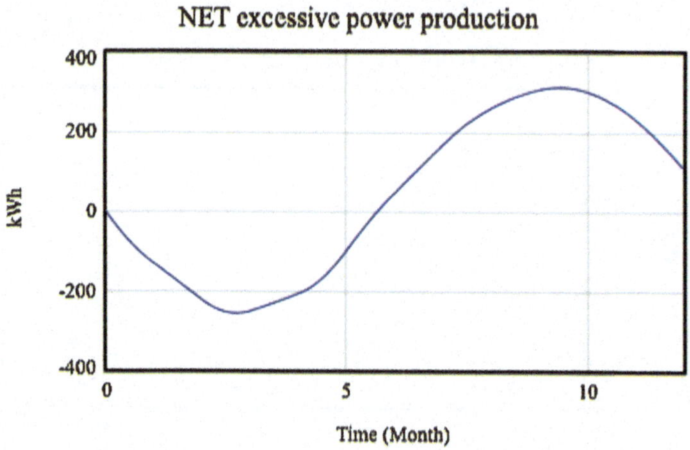

Fig. 12.7 Net excessive electric power production in 2017

The resulting accumulated monthly total solar power production in 2017 is depicted in Fig. 12.6.

Figure 12.7 shows a positive net excessive power production during the final months of the year 2017. A negative value corresponds with a deficit in self-generated electric power and results in consuming electric power from the grid. Positive values correspond with potential excessive power that can be used for electric heating.

In order to determine whether excessive net electric power can be used for electric heating, we have to distract the net excessive power production from the projected electric power need. Figure 12.8 depicts the monthly projected electric power need.

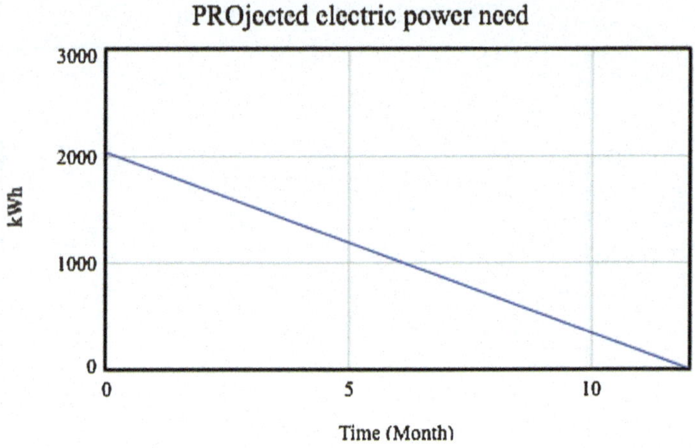

Fig. 12.8 Monthly projected electric power need

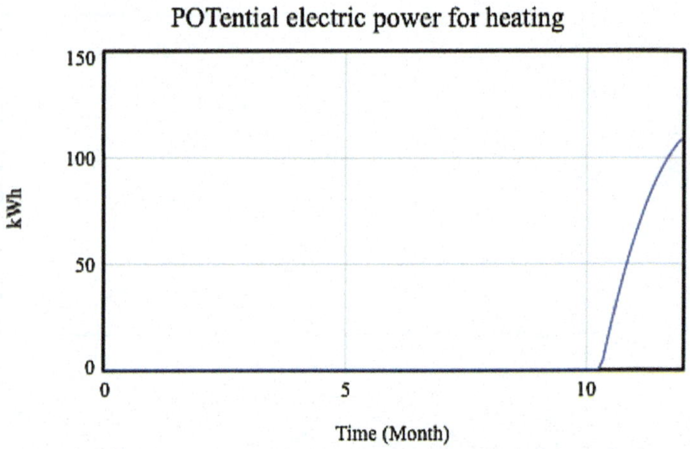

Fig. 12.9 Potential electric power that can be used for electric heating in winter

The electric power that is estimated to be used for electric heating is determined by the IF.. THEN ELSE function: IF THEN ELSE (gas consumption input rate > 100, MAX((NET excessive power production-PROjected electric power need),0),0). The resulting potential electric power for heating is shown in Fig. 12.9.

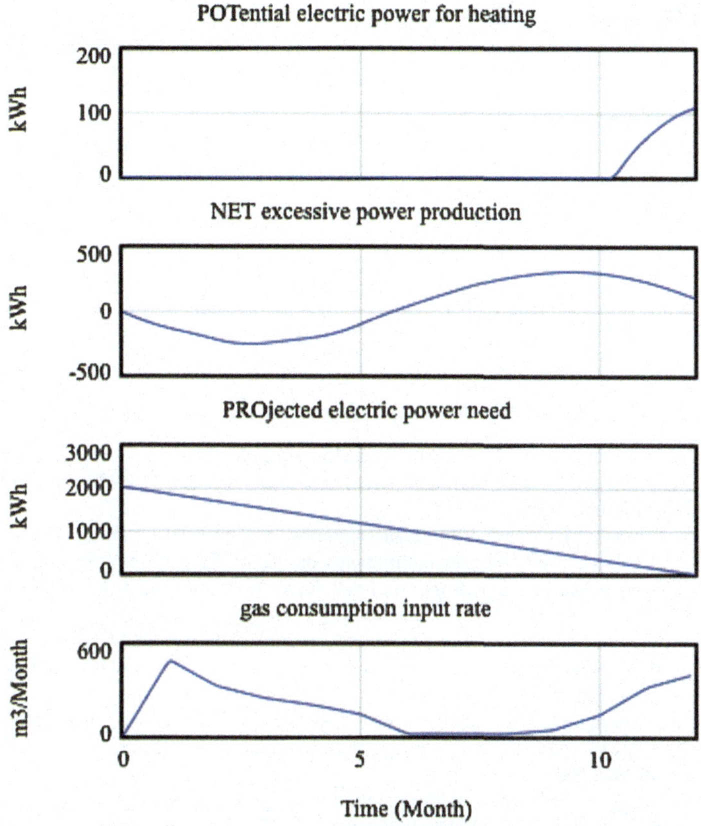

Fig. 12.10 Causal strip for the auxiliary variable "POTential electric power for heating"

Figure 12.10 shows the causal strip for the behavior of the variable "POTential electric power for heating". It depicts the evolution over time of the three input variables that were discussed above.

References

Annesi-Maesano, I., Forastiere, F., Kunzli, N., & Brunekref, B. (2007). Editorial particulate matter, science and EU policy. *The European Respiratory Journal, 29*, 428–431. https://doi.org/10.1183/09031936.00129506.

Banasik, A., Bloemhof-Ruwaard, J. M., Kanellopoulos, A., Claassen, G. D. H., & van der Vorst, J. G. A. (2016). Multi-criteria decision making approaches for green supply chains: A review. *Flexible Services and Manufacturing Journal, 26*, 1–31. https://doi.org/10.1007/s10696-016-9263-5.

Batelle Labs. (1975). *Final report on leaf burning program*. Contract 69-02-1409.

Baxter, L. (2005). Biomass-coal co-combustion: Opportunity for affordable renewable energy. *Fuel, 84*, 1295–1302.

BIR (Bureau of International Recycling). (2008). *Report on the environmental benefits of recycling*. Online: https://www.mgg-recycling.com/wp-content/uploads/2013/06/BIR_CO2_report.pdf. Accessed 08 May 2020.

Bloomberg. (2017). *Vanilla crop damaged after storm in Main Grower Madagascar*. Online: https://www.bloomberg.com/news/articles/2017-03-15/vanilla-crop-seen-damaged-by-worst-madagascar-storm-in-13-years. Accessed 08 May 2020.

Bolt, J., Timmer, M., & van Zanden, J. L. (2014). GDP per capita since 1820. In J. L. van Zanden, et al. (Eds.), *How was life?: Global well-being since 1820*. OECD Publishing. Online: https://doi.org/10.1787/9789264214262-7-en. Accessed 30 Apr 2020.

Bonaiuti, M. (2014). Degrowth: Tools for a complex analysis of the multidimensional crisis. *Capitalism Nature Socialism, 23*(1), 30–20.

Boubel, R. W., Darley, E. F., & Schuck, E. A. (1969). Emissions from burning grass stubble and straw. *Journal of the Air Pollution Control Association, 19*(7), 497–500. https://doi.org/10.1080/00022470.1969.10466517.

Boulanger, P. M., & Bréchet, T. (2005). Models for policy-making in sustainable development: The state of the art and perspectives for research. *Ecological Economics, 55*, 337–350.

Brunelli, M. (2015). *Introduction to the analytic hierarchy process*. Springer Briefs in Operations Research. ISSN 2195-0504 (electronic).

Burke, E. K., & Kendall, G. (Eds.). (2014). *Search methodologies: Introductory tutorials in optimization and decision support techniques*. New York: Springer. https://doi.org/10.1007/978-1-4614-6940-7_15.

Business Dictionary. (2020). *Definition of cost, profit and cash flow*. Online: www.businessdictionary.com. Accessed 02 May 2020.

D. Inghels, *Introduction to Modeling Sustainable Development in Business Processes*, https://doi.org/10.1007/978-3-030-58422-1

Carter, C. R., & Rogers, D. S. (2008). A framework of sustainable supply chain management: Moving toward new theory. *International Journal of Physical Distribution and Logistics Management, 38*(5), 360–387.

Caulfield, C. W., & Maj, S. P. (2001). A case for systems thinking and system dynamics. In *Proceedings of 2001 IEEE international conference on systems, man and cybernetics* (pp. 2793–2798). Tucson: IEEE.

Cesaro, A., Belgiorno, V., & Guida, M. (2015). Compost from organic solid waste: Quality assessment and European regulations for its sustainable use. *Recourses, Conservation and Recycling, 94*, 72–79.

Checkland, P. (1981). *Systems thinking, systems practice*. New York: Wiley.

Chopra, S., & Meindl, P. (2015). *Supply chain management: Strategy, planning, and operation* (6th Global ed., 528 pp.). Pearson Higher Education.

Cole, A., & Flenley, J. (2008). Modelling human population change on Easter Island far-from-equilibrium. *Quaternary International, 184*, 150–165.

Crawford, G., & Williams, C. (1985). A note on the analysis of subjective judgments matrices. *Journal of Mathematical Psychology, 29*(4), 387–405.

Currie, W., & Galliers, B. (1999). *Rethinking management information systems: An interdisciplinary perspective*. Oxford: Oxford University Press.

Curtis, L. (2002). *Biomass burning: Wood, leaves, grass, forests, crops and trash*. Burning Issues, Special Edition.

Daly, H. E. (1972). In defense of a steady-state economy. *American Journal of Agricultural Economics, 54*, 945–954.

Daly, H. E. (2008). *Ecological economics and sustainable development: Selected essays of Herman Daly* (Advances in ecological economics series, 288 pp.). Cheltenham: Edward Elgar. ISBN: 978-1-84720-988-7.

Daly, H. E. (2014). *Three limits to growth*. Online: https://ourworld.unu.edu/en/three-limits-to-growth. Accessed 01 May 2020.

Deb, K. (2009). *Multi-objective optimization using evolutionary algorithms*. Kanpur: Wiley.

Dehghanian, F., & Mansour, S. (2009). Designing sustainable recovery network of end-of-life products using genetic algorithm. *Resources, Conservation and Recycling, 53*, 559–570.

EAA – European Aluminium Association. (2013). *Environmental profile report for the European Aluminium Industry*. Online: https://european-aluminium.eu/media/1329/environmental-profile-report-for-the-european-aluminium-industry.pdf. Accessed 10 May 2020.

ECRN (Energy Research Center Netherlands). (2017). *Phyllis database for biomass and waste*. http://www.ecn.nl/phyllis/. Accessed 01 June 2017.

Elkington, J. (1994). Towards the sustainable corporation: Win-win-win business strategies for sustainable development. *California Management Review, 36*(2), 90–100.

Emmerich, M. T. M., & Deutz, A. H. (2018). A tutorial on multiobjective optimization: Fundamentals and evolutionary methods. *Natural Computing, 17*(3), 585–609.

EPA. (1986) Technical support document for residential wood combustion, EPA Document - 450/4-85-012.

EP&C. (2000). *Directive 2000/53/EC of the European Parliament and of the Council of 18 September 2000 on end-of life vehicles*. Online: https://eur-lex.europa.eu/resource.html?uri=cellar:02fa83cf-bf28-4afc-8f9f-eb201bd61813.0005.02/DOC_1&format=PDF. Accessed 01 May 2020.

EP&C. (2003). *Directive 2003/108/EC of the European Parliament and of the Council of 8 December 2003 amending Directive 2002/96/EC on waste electrical and electronic equipment (WEEE)*. Online: https://eur-lex.europa.eu/legal-ontent/EN/TXT/PDF/?uri=CELEX:32003L0108&from=nl. Accessed 01 May 2020.

EP&C. (2008). *Directive 2008/98/EC of the European Parliament and of the Council of 19 November 2008 on waste and repealing certain Directives*. Online: https://eur-lex.europa.eu/legal-content/EN/TXT/PDF/?uri=CELEX:32008L0098&from=EN. Accessed 01 May 2020.

EP&C. (2009). *Directive 2009/28/EC on the promotion of the use of energy from renewable sources and amending and subsequently repealing Directives 2001/77/EC and 2003/30/EC. L140/16, Brussels, Belgium.* Online: https://www.energy-community.org/pls/portal/docs/360177.PDF (online: June 30, 2015).

European Commission. (2018). *Report from the commission to the European Parliament and the Council – Report on the functioning of the European carbon market.* Online: https://ec.europa.eu/clima/sites/clima/files/ets/docs/com_2018_842_final_en.pdf. Accessed 19 Oct 2019.

European Commission, EC. (2008). *Handbook on estimation of external costs in the transport sector Internalisation Measures and Policies for All external Cost of Transport (IMPACT) Version 1.1, Delft, CE, 2008.* Online: http://ec.europa.eu/transport/themes/sustainable/doc/2008_costs_handbook.pdf. Accessed 08 Nov 2015.

European Parliament. (2009). *Regulation (EC) No 443/2009 of the European Parliament and of the Council of 24 April 2009 setting emission performance standards for new passenger cars as part of the Community's integrated approach to reduce CO2 emissions from light-duty vehicles.* Online: http://eur-lex.europa.eu/LexUriServ/LexUriServ.do?uri=OJ:L:2009:140:0001:0015:EN:PDF. Accessed 06 July 2014.

Forrester, J. W. (1989). *The beginning of system dynamics.* Banquet Talk at the international meeting of the System Dynamics Society Stuttgart, Germany. Online: http://web.mit.edu/sysdyn/sd-intro/D-4165-1.pdf. Accessed 01 Feb 2020.

Fowler, P., Krajacic, G., Loncar, D., & Duic, N. (2009). Modelling the energy potential of biomass-H2RES. *International Journal of Hydrogen Energy, 34*, 7027–7040.

Friedman, L., & Calabrese, E. (1977). The health implications of open leaf burning. *Reviews in Environmental Health, 2*(4), 257–283.

Global Footprint Network. (2019). *Earth overshoot day 2019 is July 29th, the earliest ever.* Online: https://www.footprintnetwork.org/2019/06/26/press-release-june-2019-earth-overshoot-day/. Accessed 30 Dec 2019.

Goedkoop, M., & Spriensma, R. (2001). *Eco-indicator 99-a damaged oriented method for life cycle assessment.* Methodology report 3rd ed. Online: www.pre.nl. Accessed 10 Oct 2011.

Grant, T. (2003). Maximizing resource recovery from waste streams. *Environmental Progress, 22* (4), 250–254. https://doi.org/10.1002/ep.670220413.

Gungor, A., & Gupta, S. M. (1999). Issues in environmentally conscious manufacturing and product recovery: A survey. *Computers & Industrial Engineering, 36*, 811–853.

Gwozdz, W., Steensen Nielsen, K., & Müller, T. (2017). An environmental perspective on clothing consumption: Consumer segments and their behavioral patterns. *Sustainability, 9*, 762. https://doi.org/10.3390/su9050762.

Halog, A., & Manik, Y. (2011). Advancing integrated systems modelling framework for life cycle sustainability assessment. *Sustainability, 3*, 469.

Hardin, G. (1968). The tragedy of the commons. *Science, New Series, 162*(3859), 1243–1248. Published by: American Association for the Advancement of Science. http://www.jstor.org/stable/1724745

Hediger, W. (1999). Reconciling "weak" and "strong" sustainability. *International Journal of Socio Economics, 26*(7/8/9), 1120–1144.

Heijungs, R., Huppes, G., & Guinée, J. B. (2010). Life cycle assessment and sustainability analysis of products, materials and technologies: Toward a scientific framework for sustainability life cycle analysis. *Polymer Degradation and Stability, 95*, 422–428.

Hillier, F. S., & Lieberman, G. J. (2001). *Introduction to operations research.* Dubuque: McGraw-Hill Higher Education.

IEA. (2019). *Global EV outlook 2019.* Online: https://www.iea.org/reports/global-ev-outlook-2019. Accessed 16 May 2020.

Ilgin, M. A., & Gupta, S. M. (2010). Environmentally conscious manufacturing and product recovery (ECMPRO): A review of the state of the art. *Journal of Environmental Management, 91*, 563–591.

Inghels, D., Dullaert, W., Raa, B., & Walther, G. (2016a). Influence of composition, amount and life span of passenger cars on end-of-life vehicles waste in Belgium: A system dynamics approach. *Transportation Research Part A, 81*, 80–104.

Inghels, D., Dullaert, W., & Bloemhof, J. (2016b). A model for improving sustainable green waste recovery. *Resources, Conservation and Recycling, 110*, 61–73. https://doi.org/10.1016/j.resconrec.2016.03.013.

Inghels, D., Dullaert, W., & Vigo, D. (2016c). A service network design model for multimodal municipal solid waste transport. *European Journal of Operational Research*. https://doi.org/10.1016/j.ejor.2016.03.036.

Inghels, D., Dullaert, W., Aghezzaf, E. H., & Heijungs, R. (2019). Towards optimal trade-offs between material and energy recovery for green waste. *Waste Management, 93*, 100–111.

IPCC. (2020). *Reports*. Online: https://www.ipcc.ch/2019/. Accessed 01 Feb 2020.

ISO 14040. (2006). *Environmental management – Life cycle assessment -principles and framework*. Online: www.iso.org

ISO 14044. (2006). *Environmental management – Life cycle assessment-requirements and guidelines*. Online: www.iso.org

ISO 1421. (2016). *Environmental labels and declarations – Self-declared environmental claims (Type II environmental labelling)*. Online: https://www.iso.org/standard/66652.html

Jacquemin, L., Pontalier, P. Y., & Sablayrolles, C. (2012). Life cycle assessment (LCA) applied to the process industry: A review. *International Journal of Life Cycle Assessment, 17*, 1028–1041. https://doi.org/10.1007/s11367-012-0432-9.

Jayal, A. D., Badurdeen, F., Dillon, O. W., & Jawahir, I. S. (2010). Sustainable manufacturing: Modelling and optimization challenges at the product, process and system levels. *CIRP Journal of Manufacturing Science and Technology, 2*(3), 144–152.

Jeswani, H. K., Azapagic, A., Schepelmann, P., & Ritthoff, M. (2010). Options for broadening and deepening the LCA approaches. *Journal of Cleaner Production, 18*, 120–127.

Kauffmann, R. G., & Adams, J. (2019, January 17–19). *International supply chain sustainability*. Fifteenth international conference on environmental, cultural, economic, and social sustainability, Vancouver, BC, Canada.

Kibira, D., Jain, S., & McLean, C. (2009). *A system dynamics modelling framework for sustainable manufacturing*. Online: http://www.systemdynamics.org/conferences/2009/proceed/papers/P1285.pdf. Accessed 01 Mar 2014.

Kleindorfer, P. R., Singhal, K., & Van Wassenhove, L. (2005). Sustainable Operations Management. *Production and Operations Management, 14*(4), 482–492.

Klöpffer, W. (2008). Life cycle sustainability assessment of products (with Comments by Helias A. Udo de Haes, p. 95). *International Journal of Life Cycle Assessment, 13*(2), 89–95. https://doi.org/10.1065/lca2008.02.376.

Kuhn, T. S. (1962). *The structure of scientific revolutions*. International Encyclopedia of Unified Science. ISBN: 0-226-45803-2 (clothbound); 0-226-45804-0 (paperbound). Online: https://projektintegracija.pravo.hr/_download/repository/Kuhn_Structure_of_Scientific_Revolutions.pdf

Lethonen, M. (2004). The environmental-social interface of sustainable development: Capabilities, social capital, and institutions. *Ecological Economics, 49*, 199–214.

Levi Strauss & CO. (2015). *The life cycle of a jean; understanding the environmental impact of a pair of Levi's® 501® jeans*. Online: http://levistrauss.com/wp-content/uploads/2015/03/Full-LCA-Results-Deck-FINAL.pdf. Accessed 08 May 2020.

Life Cycle Initiative. (2020). *Benefits of life cycle approaches*. Online: https://www.lifecycleinitiative.org/starting-life-cycle-thinking/benefits/. Accessed 01 May 2020.

Lipkin, L., & Smith, D. (2004). Logistic growth model – Background: Logistic modeling. *Convergence*. Online (21/07/2018): https://www.maa.org/press/periodicals/loci/joma/logistic-growth-model-background-logistic-modeling

Martinez-Alier, J., Munda, G., & O'Neill, J. (1998). Weak comparability of values as a foundation for ecological economics. *Ecological Economics, 26*, 277–286.

Mavrotas, G. (2007). *Generation of efficient solutions in multi-objective mathematical programming problems using GAMS*. Effective implementation of the ε-constraint method, available on www.GAMS.com. Accessed 21 May 2015.

Mavrotas, G. (2009). Effective implementation of the ε-constraint method in multi-objective mathematical programming problems. *Applied Mathematics and Computation, 213*, 455–465.

Mavrotas, G., & Florios, K. (2013). An improved version of the augmented ε-constraint method (AUGMECON2) for finding the exact pareto set in multi-objective integer programming problems. *Applied Mathematics and Computation, 219*(18), 9652–9669.

McKendry, P. (2002a). Energy production from biomass (part 1): Overview of biomass. *Bioresource Technology, 83*(1), 37–46.

McKendry, P. (2002b). Energy production from biomass (part 2): Conversion technologies. *Bioresource Technology, 83*(1), 47–54.

Meadows, D. H., Meadows, D. L., Randers, J., & Behrens, W. W., III. (1972). *The limits to growth: A report for the club of Rome's project on the predicament of mankind*. New York: Potomac Associates.

Mirchi, A., Madani, K., Watkins, D., Jr., & Ahmad, S. (2012). Synthesis of system dynamics tools for holistic conceptualization of water resources problems. *Water Resources Management, 26*, 2421–2442. https://doi.org/10.1007/s11269-012-0024-2.

MIT. (1999). *Generic structures: Exponential material delays*. Online: https://ocw.mit.edu/courses/sloan-school-of-management/15-988-system-dynamics-self-study-fall-1998-spring-1999/readings/generic5.pdf. Accessed 30 Apr 2020.

Moldavska, A., & Welo, T. (2016). Development of manufacturing sustainability assessment using systems thinking. *Sustainability, 8*(1), 5. https://doi.org/10.3390/su8010005.

Morrissey, A. J., & Browne, J. (2004). Waste management models and their application to sustainable waste management. *Waste Management, 24*, 297–308.

Mujkić, Z., Kraslawski, A., & Qorri, A. (2018). Sustainability and optimization of supply chains: A literature review. *Operations and Supply Chain Management, 11*(4), 186–199.

Munda, G. (2004). Social multi-criteria evaluation: Methodological foundations and operational consequences. *European Journal of Operational Research, 158*, 662–677.

Munda, G. (2005). Chapter 23: Multiple criteria decision analysis: State of the art surveys. In J. Figueira, S. Greco, & M. Ehrgott (Eds.), *Multiple criteria decision analysis and sustainable development*. Boston: Springer. Online: http://citeseerx.ist.psu.edu/viewdoc/download?doi=10.1.1.136.9170&rep=rep1&type=pdf. Accessed 14 Feb 2016.

Narzisi, G. (2008). *Classic methods for multi-objective optimization*. Courant Institute of Mathematical Sciences New York University. Online: https://www.semanticscholar.org/paper/Classic-Methods-for-Multi-Objective-Optimization-Narzisi/0bd08c5bb8ee851d5ad0ca8d4a8eb820d007409f. Accessed 01 June 2020.

Nidumolu, R., Prahalad, C. K., & Rangaswami, M. R. (2009). Why sustainability is now the key driver of innovation. *Harvard Business Review*. Online: https://hbr.org/2009/09/why-sustainability-is-now-the-key-driver-of-innovation. Accessed 12 Oct 2019.

OECD. (2017). *ITF transport outlook 2017*. Online: http://www.oecd-library.org/docserver/download/e979b24d-en.pdf?expires=1504418158&id=id&accname=guest&checksum=63E2C96A857B0E36EEF829275424B8CE. Accessed 20 May 2019.

Oglethorpe, D. (2010). Optimizing economic, environmental, and social objectives: A goal programming approach in the food sector. *Environment and Planning A, 42*, 1239–1254.

OVAM. (2009). Economische marktanalyse voor een duurzame verwerking van (deelstromen) van groen- en gft-afval met voorstel van beleidsaanbevelingen [Economic market analysis for the sustainable treatment of green and VFG waste with proposal for policy recommandations], D/2009/5024/84. Mechelen, Belgium.

Panteia. (2013). *Contribution to impact assessment of measures for reducing of emissions in Inland navigation*. Reference R20130023/C0019200/RLE/LJO. Online: http://ec.europa.eu/transport/modes/inland/studies/doc/2013-06-03-contribution-to-impact-assessment-of-measures-for-reducing-emissions-of-inland-navigation.pdf. Accessed 11 Nov 2013.

Peeters, J. (red) (2015). *Veerkracht en burgerschap*. EPO.
Pick, D., Dieterich, M., & Heintschel, S. (2012). Biogas production potential from economically usable green waste. *International Journal of Sustainable Development and World Ecology, 4* (4), 682–702. https://doi.org/10.3390/su4040682.
Piekarski, C. M., da Luz, L. M., Zocche, L., & de Francisco, A. C. (2013). Life cycle assessment as entrepreneurial tool for business management and green innovations. *Journal of Technology Management & Innovation, 8*(1), 44–53.
Piketty, T. (2014). *Capital in the twenty-first century*. Cambridge: Harvard University Press.
Pope, J., Annandale, D., & Morrison-Saunders, A. (2004). Conceptualizing sustainability assessment. *Environmental Impact Assessment Review, 24*, 595–616.
Rees, W. E. (1988). A role for environmental impact assessment in achieving sustainable development. *Environmental Impact Assessment Review, 8*, 273–291.
Rifkin, J. (2010). *The empathic civilization*. Cambridge: Polity Press.
Roser, M. (2020). *Economic growth*. Published online at OurWorldInData.org. Retrieved from https://ourworldindata.org/economic-growth. Accessed 17 Apr 2020.
S&P Global Platts. (2018). *EU carbon market stages comeback in 2018*. Online: https://blogs.platts.com/2018/11/26/insight-eu-carbon-market-comeback-2018/. Accessed 18 May 2020.
Saaty, T. L. (1980). *The analytic hierarchy process: Planning, priority setting, resource allocation*. New York: McGraw Hill.
Saaty, T. L. (2006). *The analytic network process*. Archive of SID. Online: www.SID.ir. Accessed 10 Dec 2011.
Saaty, T. L. (2008). Decision making with the analytic hierarchy process. *International Journal of Services Sciences, 1*(1), 83–98.
Saaty, T. L., & Vargas, L. (1991). *Prediction, projection and forecasting: Applications of the analytic hierarchy process in economics, finance, politics, games and sports* (254 p.). New York: Kluwer Academic Publishers/Springer. ISBN 978-94-015-7954-4.
Schmelzer, M. (2015). The growth paradigm: History, hegemony, and the contested making of economic growthmanship. *Ecological Economics, 118*, 262–271.
SenterNovem. (2008). *Verwerking "Gescheiden ingezameld groenafval" – ontwerpversie* [Treatment of separated collected green waste- draft version]. Online: http://www.lap2.nl/sn_documents/files/04%20LCA-LAP2/LCA_groenafval_ontwerpversie.pdf. Accessed 10 Sept 2012.
Seuring, S. (2013). A review of modelling approaches for sustainable supply chain management. *Decision Support Systems, 54*, 1513–1520.
Seuring, S., & Goldbach, M. (2002). *Cost management in supply chains* (445 p.). Heidelberg: Physica-Verlag. ISBN 3-7908-1500-4.
Seuring, S., & Müller, M. (2008). From a literature review to a conceptual framework for sustainable supply chain management. *Journal of Cleaner Production, 16*, 1699–1710.
Sharma, B., Ingalls, R. G., Jones, C. L., & Khanchi, A. (2013). Biomass supply chain design and analysis: Basis, overview, modelling, challenges, and future. *Renewable & Sustainable Energy Reviews, 24*, 608–627.
Slaper, T. F. (2011). The triple bottom line: What is it and how does it work? *Indiana Business Review*. Online: https://www.ibrc.indiana.edu/ibr/2011/spring/pdfs/article2.pdf
Smith, A. (1776). *An inquiry into the nature and causes of the wealth of nations*. Oxford: OUP. 2008.
Sterling, S. (2004). Higher education, sustainability, and the role of systemic learning. In P. Corcoran & A. Wals (Eds.), *Higher education and the challenge of sustainability. Problematics, promise and practice* (pp. 59–70). Dordrecht: Kluwer Academic Publishers.
Sterman, J. D. (2000). *Business dynamics: System thinking and modelling for a complex world*. Indianapolis: McGraw-Hill.
Stockholm Resilience Centre. (2020). *Planetary boundaries research*. Online: www.stockholmresilience.org. Accessed 20 May 2020.

Tajbakhsh, A., & Hassini, E. (2015). Performance measurement of sustainable supply chains: A review and research questions. *International Journal of Productivity and Performance Management, 64*(6), 744–783. https://doi.org/10.1108/ijppm-03-2013-0056.

Tang, V., & Vijay, S. (2001, Fall). *System dynamics origins, development, and future prospects of a method*. ESD.83. Research seminar in engineering systems. Online: web.mit.edu/esd.83/www/.../System%20Dynamics%20final.doc. Accessed 23 June 2018.

Thuesen, G. J., & Fabrycky, W. J. (1993). *Engineering economy*. Englewood Cliffs: Prentice Hall.

Towill, D. R. (1996). Industrial dynamics modelling of supply chains. *Logistics Information Management, 9*(4), 43–56.

Ucak, A. (2015). Adam Smith: The inspirer of modern growth theories. World conference on technology, innovation and entrepreneurship. *Procedia – Social and Behavioral Sciences, 195*, 663–672.

United Nations (UN). (1992). *Agenda 21, United Nations conference on environment & development Rio de Janerio, Brazil*. Online: https://sustainabledevelopment.un.org/content/documents/Agenda21.pdf. Accessed 16 Jan 2016.

Vadenbo, C., Hellweg, S., & Guillén-Gosálbez, G. (2014). Multi-objective optimization of waste and resource management in industrial networks – Part I: Model description. *Resources, Conservation and Recycling, 89*, 52–63.

Von Bertalanffy, L. (1968). *General system theory. Foundations, development, applications*. New York: George Braziller.

Wilkinson, R., & Pickett, K. (2011). *The spirit level: Why greater equality makes societies stronger*. Reprint edition. New York: Bloomsbury Press.

Wimmer, W., Züst, R., & Lee, K. M. (2004). Chapter 2: Life cycle assessment. In *Ecodesign implementation* (Volume 6 of the series alliance for global sustainability book series, pp. 19–61). Berlin: Springer. Online: http://link.springer.com.vu-nl.idm.oclc.org/chapter/10.1007/978-1-4020-3071-0_2

Woodward, D. G. (1997). Life cycle costing--theory, information acquisition and application. *International Journal of Project Management, 15*(6), 335–344.

World Commission on Environment and Development (WCED). (1987). *Our common future*. New York: Oxford University Press.

The manufacturer's authorised representative in the EU is Springer Nature Customer Service Centre GmbH, Europaplatz 3, 69115 Heidelberg, Germany. If you have any concerns regarding our products, please contact ProductSafety@springernature.com

Printed and bound by CPI Group (UK) Ltd, Croydon, CR0 4YY
30/09/2026
02225445-0003